COPPER PORPHYRIES

Other Miller Freeman Publications books for the mining and minerals processing industries:

Minerals Transportation Volume 1: Proceedings of the first International Symposium on Transport and Handling of Minerals, Vancouver, British Columbia, Canada, October 1971

Minerals Transportation Volume II: Proceedings of the second International Symposium on the Transport and Handling of Minerals, Rotterdam, Netherlands, October 1973

Tailing Disposal Today: Proceedings of the first International Symposium on Tailing Disposal, Tucson, Arizona, U.S.A., October-November 1972

World Mines Register: International directory of active mining companies, operations, and key personnel

*Waste Production and Disposal in the Mining, Milling, and Metallurgical Industries by Roy E. Williams

*International Glossary of Mining, Processing and Geological Terms: Revised edition of over 10,000 terms and phrases. Translated in five languages: English, French, German, Spanish, and Swedish, with four separate bilingual indices.

*Scheduled for 1975 publication

COPPER PORPHYRIES

ALEXANDER SUTULOV

Miller Freeman Publications, Inc.

BY THE SAME AUTHOR:
  Molibdeno, Santiago de Chile, 1962
  Proceso de Segregacion, Concepcion, 1962
  Flotacion de Minerals, Concepcion, 1963
  Proceso L.P.F., Concepcion, 1963
  Molybdenum Extractive Metallurgy, 1965
  Mineralurgia Latinoamericana, 1966
  Copper Production in Russia, 1967
  Molybdenum and Rhenium Recovery
      from Porphyry Coppers, 1970
  The Soviet Challenge in Base Metals, 1971
  Minerals in World Affairs, 1972
  Mineral Resources and the
      Economy of the USSR, 1973

© Copyright 1975 (Index) by MILLER FREEMAN PUBLICATIONS, Inc. 500 Howard Street, San Francisco, California 94105 USA. Printed in the United States of America.

Library of Congress Catalog Card Number 73-93695
International Standard Book Number 0-87930-028-0

All rights reserved. No part of this work covered by the copyrights hereon may be reproduced or copied in any form or by any means—graphic, electronic, or mechanical, including photocopying, recording, taping, or information and retrieval systems—without written permission from the publisher.

First printing January 1974
Second printing April 1975

Table of Contents

PREFACE 5

Chapter One : INTRODUCTION 7

Chapter Two : PORPHYRY COPPER DEPOSITS 13
Definition; Porphyry Model; Genesis of Copper Deposits; Age of Copper Porphyries; Economic Importance of Copper Porphyries; Bibliography.

Chapter Three : ECONOMIC GEOGRAPHY OF
PORPHYRY COPPERS 27
Latin America; Chilean Porphyries; Peruvian Porphyries; Other Latin American Porphyries; Mexican Deposits; United States Porphyries; British Columbian Deposits; Pacific Fire Belt; Alpide Porphyries; Bibliography.

Chapter Four : METAL PRODUCTION FROM
PORPHYRY COPPERS 53
General; U. S. Production; Chilean Production; Peruvian Production; Canadian Production; Pacific Belt Porphyries; Communist Bloc Porphyries; Projects Under Development; Bibliography.

Chapter Five : METALLURGY OF COPPER
PORPHYRIES 99
Crushing and Grinding; Flotation Technology; Rougher Flotation; Collectors; Frothers; Conditioning Reagents; Crowding Out Effect and Lag; Relationship between Mineralization and Reagents Use; Cleaner Flotation; Special Processes; LPF Process; Copper Hydrometallurgy; Solvent Extraction; Investment and Production Costs; Bibliography.

Chapter Six : MOLYBDENITE AND RHENIUM
BY-PRODUCT RECOVERY 141
Molybdenite Occurrences and Properties; Problems of Molybdenite Separation; Thermal Depression; Copper Depression by Chemical Means; Nokes Reagents; The Russian Sulfidization Process; Western Sulfidization Processes; Molybdenite Flotation; Rhenium in Porphyry Coppers; Production Technology; Investment and Production Costs; Bibliography.

Chapter Seven : SMELTING AND REFINING
OF COPPER 171
Roasting of Copper Concentrates; Smelting of Calcines and Concentrates; Converting; Alternate Solutions; Electric Smelting; Flash Smelting; Noranda Process; WOCRA Process; Mitsubishi Process; Current Practices and Tendencies; Chemical Smelting; Copper Refining; Refining Process; Smelting and Refining Costs; Total Costs of Copper Production; Bibliography.

INDEX 201

Preface

Copper porphyries are a typical twentieth-century phenomenon. Known for many years as very low-grade deposits, they acquired prominence at the beginning of this century when an American engineer, Daniel Jackling, proved their economic value by operating them on a large scale. Gradually they were transformed into the backbone of our copper supplies, and in this century will provide us with some 50 to 60 percent of all new copper consumed. They will close the gap in our copper supplies by gradually fading out into very low-grade large masses of overburden, mine rejects, dumps and tailings disposals or simply remain in situ in sub-marginal grades until such time as large-scale and very cheap methods, or gradual increases in copper prices, make them commercially feasible for exploitation by chemical and atomic mining methods toward the end of this century or the beginning of the next.

In this book the author has tried to give a general picture of the importance of porphyry coppers as a natural resource, both geologically and metallurgically, rather than to engage in a detailed discussion of any particular technical subject. This did not prevent him from discussing in some detail the subject of contemporary technology and future projections, particularly if these were critically connected with costs, ecological problems or other important matters. By doing so, national borders were overcome and what once appeared as a typical North American art now had been converted into a world-wide phenomenon, including the Communist World and the Third World.

The exploitation of porphyry coppers began in the United States at the very beginning of this century. It spread rapidly to Latin America, first to Chile and Mexico, and later to Peru and other parts. By the Second World War it had reached Russia, and after the war spread into the Pacific area and British Columbia. So far, only one third of the known porphyry copper ore reserves have been taken out of the earth's crust, and before the end of this century twice again as much copper will be extracted. But, as shown in this book, this will not be the end of our porphyry copper reserves. As a matter of fact, this may just be the beginning of a new and great chapter in the history of the red metal. It is the author's intention to reveal in this book what the immediate future holds for us concerning this fascinating subject.

Salt Lake City

November 7, 1973

CHAPTER ONE

Introduction

Since the beginning of our civilization, man has discovered in the earth's crust, orebodies containing about 600 million tons* of metallic copper and he has consumed about one third of these reserves. His appetite for copper has grown in a rather impressive way: 1 million tons in the first 60 centuries or so of our civilized life;[1] 10.3 million tons throughout the 19th century, when major discoveries of electrical energy uses were made; and finally at least some 180 million tons in the first 73 years of this century.

It is on record that while in 1800 man produced and consumed some 18,000 tpy of copper, his production and consumption by the end of that century surpassed 500,000 tpy. It rose to over 2,000,000 tpy by the end of the twenties; 5,000,000 tpy in the early sixties; and today is estimated at 9,000,000 tpy of new copper.

If this trend continues, with a 4 to 5 percent annual growth rate as it has in the last two decades, then by 1980 our copper consumption will be 12 million tons per year; by 1990 about 19 million tons; and at least 30 million tons in the year 2000.

It is necessary to observe that the known copper reserves, evaluated at this moment at some 420 million tons, and the estimated 100 million tons of copper in use, which supply most of the scrap for secondary copper, are sufficient for roughly 40 to 50 years of our needs at the present rate of consumption, and will not even last until the end of this century at the present accelerated rate of consumption.

Surprisingly, this situation really does not seriously bother anyone and copper is still considered to be a relatively abundant and readily available raw material, which can be profitably and effectively gained from the earth's crust at a reasonable cost.

---

* Throughout this book short tons of 2,000 lbs. will be used.

TABLE 1.1
WORLD COPPER RESERVES
(in millions of short tons)

	Recognized			Hypothetical
	Porphyry	Other	Total	
United States	86	14	100	100
Canada	15	10	25	50
Mexico	12	6	18	20
Central America	6	1	7	5
Chile	86	4	90	45
Peru	21	4	25	50
Other South America	9	1	10	20
Western Hemisphere	235	40	275	290
Europe	4	21	25	20
USSR	6	33	39	50
Africa	—	53	53	80
Middle East	4	—	4	20
Oceania & Japan	11	10	21	50
Australia	—	3	3	10
Eastern Hemisphere	25	120	145	230
WORLD TOTAL	260	160	420	520

The reason for this is that the known copper resources have a great potential for growth, provided effective technology and intelligent management can be exercised in the exploitation of low-grade copper deposits, most of which belong to the porphyry type.

One of the leading geologists and specialists in porphyry coppers, David Lowell, estimates[2] that within the first mile of the earth's crust is contained some 3,000 trillion tons of metallic copper, a quantity sufficient to supply our present demand for another 330,000,000 years at the present rate of consumption.

While such a quantity of copper could satisfy our necessities eternally, even by means of the effective circulation of a great quantity of copper in pool, it is obvious that not all of this copper can be considered a resource because of the high degree of its dispersion. For it would take a very special technology, probably similar to chemical analytical methods, to recover copper from ores which contain only 70 parts per million of metal, and then, of course, it would be very difficult to mine out the first mile of our earth's crust, particularly taking into consideration that this would take much of our land territory under the sea.

Thus for all practical purposes, we should consider for copper recovery only those parts of the earth's crust where the copper concentration is sufficiently high to warrant economic extraction. Presently this cut-off grade is considered to be around 0.4 percent copper, but cases are known where even lower grade ores are economically processed.

By taking an arbitrary 0.25% Cu cut-off grade, our minable copper resources would be reduced to 1/300,000 of the copper content of the first mile of the earth's crust, i.e., about 10,000,000 tons of metallic copper, affirms Lowell.

This obviously radically changes our copper resources situation, since it increases our known copper reserves some 25 times.

This exercise in copper inventory was undertaken with the purpose of demonstrating the importance and great future potential of the copper porphyries as the backbone of our copper supply. If not for this type of orebody, our available copper supplies would have been severely limited by the turn of the century, when 4 percent copper ores were considered to be the average acceptable material for treatment; and without any doubt the role of copper as a modern technological material would be severely limited simply because of its unavailability in sufficient quantities to satisfy demand.

We should probably recall that this critical situation — shortage of readily available high-grade ores — cost Chile its leadership in world copper production in the last quarter of the 19th century. It is to the credit of an American metallurgist, Daniel C. Jackling, that he recognized the possibility of profitable exploitation of low-grade copper porphyries at the turn of the century and thus avoided a crisis in the supply of this relatively cheap technological and strategic metal. Incidentally, the fact that the first copper porphyries were put into production in the United States greatly contributed to the conversion of the United States into the leading copper producer and the greatest industrial power of the world.

The Jackling idea was relatively simple: by using large-scale and technologically effective methods, lower grade ores could be processed economically and the lesser profits on leaner ore converted into important income because of the larger scale of operations and greater volume of material treated and produced. In technical jargon, fixed costs go down with the increased scale of operations, making lower profit margins interesting because of the scale of operations used, a concept widely accepted and applied today by copper industry.

In his design, Jackling took full advantage of the modern mining technology then recently available, using steam shovels and cheap transportation; he made effective use of gravity concentration and swiftly changed to flotation as soon as this method of concentration became technically available; but more importantly, his first designed mill at Bingham, Utah, in 1906 was a 2,000 tpd operation rather than a typical plant for that time of 200 to 300 tpd, and it grew immediately into a 6,000 tpd operation after the initial success was proved.

The Jackling ideas rapidly materialized into new projects, both in the United States and abroad. Development of open pit mining, effective and cheap materials moving methods, selective concentration methods, recovery of by-products and overall reduction of investment and processing costs contributed greatly to the abundant supply of cheap copper, and the mill feed, which in the beginning of this century averaged 2 percent copper, could subsequently be reduced to 1 percent two decades ago, 0.5 percent one decade ago and to even lower levels today. Simultaneously, the scale of operations grew from 6,000 tpd to 15,000 tpd and 20,000 tpd in the mid-thirties; 30,000 to 40,000 tpd in the fifties and sixties, and to greater than 60,000 tpd today. Even now there is under consideration a project where a mill capacity of 176,000 tpd would be provided with an overall investment of more than $500 million, a figure 100 times in excess of the initial Guggenheim investment in the Bingham project.

The record of discovery of copper deposits in the last quarter of a century indicates that approximately two out of each three tons of newly discovered copper belong to the porphyry type orebodies. These deposits with enormous masses of intrusive rock mineralized by finely disseminated copper values were transformed into the backbone of our copper supply. Fully 40 percent of the copper mined in this century and 45 percent of the presently mined copper comes from porphyry coppers. Their share in future supplies will more than likely be even more important because about 62 percent of the presently known copper reserves are locked in this type of orebody, and geological evidence indicates that this proportion may be maintained in future discoveries as well. The relatively well explored areas of Arizona and the southwestern states still offer rather challenging prospects in new copper discoveries, while the less explored areas of British Columbia, Alaska, Kazakhstan and particularly Latin America and the Pacific Fire Belt are definitely the most promising prospects for additional copper reserves. Experts believe,[2] for instance, that British Columbia alone may have another 100 copper porphyries still hidden in her territory. With this in mind it is difficult to escape the impression that the South American Andean Area or for that matter the whole southern and western part of the Pacific Fire Belt have even greater potentialities if a sufficiently intensive search for copper is made.

If we take into consideration that an average porphyry copper orebody contains about 1 million tons of metallic copper, and that many of these have 3 to 5 million tons of copper reserves, then it is obvious that literally several billion tons of metallic copper are still waiting for discovery and extraction for our expanding demand. There will be no prospect of a "copper crisis" like our present "energy crisis" if opportune prospection and exploration operations are carried out and if sufficient outlets of capital are provided for development of efficient technology and construction of plants. Man's abilities to use copper as a first-class technological material are peculiarly related to his intelligence and willingness to exercise rational management of economic and technological problems, for there will be no shortage of copper as a natural resource, at least in the foreseeable future.

It is important to note that copper porphyries, similar to their past and present history, will also exercise this significant role in the future, and thus reflect the importance of our technical, economic and political decisions. And there will be a need for plenty of them in view of the fact that these deposits are located principally in economically underdeveloped and politically sensitive areas.

Bibliography for Chapter One

1. J. Newton & C. L. Wilson: Metallurgy of Copper, John Willey & Sons, N. Y., 1942.

2. J. David Lowell: Copper Resources in 1970 — The 1970 Jackling Award Lecture, Mining Engineering, April 1970, pp. 67–73.

Fig. 0.2 The oldest porphyry copper mine in the world at Bingham, Utah. (Photograph by Don Green, courtesy of Kennecott Research Center.)

CHAPTER TWO

Porphyry Copper Deposits

Definition

It would be logical to start this chapter, or even this book for that matter, with a precise and short definition of copper porphyries. This is almost impossible because so far this term has had different meanings to many people, and is used not only in its geological sense, but in economics and engineering as well.

Geologically, copper porphyry means a disseminated copper mineralization in an acid or medium acid igneous porphyritic rock. But as Schwartz,[1] Parson,[2] Titley[3] and many other authorities have pointed out, such a definition today would be incomplete, and a concise one which would incorporate other aspects is difficult. In his glossary, H. E. McKinstry[4] gives the following definition of a porphyry copper: a deposit in which the copper bearing minerals occur in disseminated grains or veinlets throughout a great volume of rock. The rock need not be porphyry. Principal characteristics: large tonnage and relatively low-grade.

Bateman[5] and Parson[2] stated that similar characteristics of the porphyry copper deposits can be summarized as follows:

1. Low-grade
2. Large tonnage
3. Association with stocklike intrusions of monzonite porphyries
4. Disseminated replacements in porphyry or intruded shists
5. Blanket shape and greater horizontal than vertical dimensions
6. Similar primary mineralogy
7. Intense seritization and, in places, silicification
8. Overlaid by leached cappings

9. More or less developed supergene enrichment
10. Susceptibility to large-scale low-cost methods of mining

Bateman noted, however,[5] that some deposits classified as porphyry coppers show differences in host rocks, shape, size, tenor, grade of oxidation and degree of supergene enrichment. Precisely, these differences will be important to our study of porphyry copper deposits, since they determine the mining of an orebody, the particular flowsheet and metallurgical treatment used for recovery of metal values and the whole concept of ore processing.

More recently, Lowell[6] characterized porphyry copper deposits as "low-grade, roughly equidimensional, disseminated deposits which contain chalcopyrite, pyrite and at least trace amounts of molybdenite, silver and gold, and which sometimes contain chalcocite and bornite. The deposits tend to have either vertical-cylindrical or flat-disc shape and are hypogene, hydrothermal deposits always related to intrusive rocks, including porphyry rock units. Mineralization can occur in either the host intrusive or wall rocks."

Porphyry Model

To generalize this definition, Lowell and Guilbert recently made a comparative study of 27 major porphyry copper deposits, and following is the description of this "typical model":[7]

"It is emplaced in late Cretaceous sediments and metasediments and is associated with a Laramide (65 million years) quartz monzonite stock. Its host intrusive rock is elongate-irregular, 4,000 x 6,000 feet in outcrop, and is progressively differentiated from quartz diorite to quartz monzonite in composition. The host is more like a stock than a dike and is controlled by regional-scale faulting. The body is oval to pipe-like, with dimensions of 3,500 x 6,000 feet and gradational boundaries. Seventy percent of 140 million tons of ore occurs in igneous host rocks, 30 percent in preore rocks. Metal values include 0.45% hypogene Cu with 0.35% supergene Cu, and 0.015% Mo. Alteration is zoned from potassic at the core (an earliest) outward through phyllic (quartz-sericite-pyrite), argillic, (quartz-kaolin-montmorillonite), and propylitic (epidote-calcite-chlorite), the propylitic zone extending 2,500 feet beyond the copper zone. Over the same interval, sulfide species vary from chalcopyrite-molybdenite-pyrite through successive assemblages to an assemblage of galena-sphalerite with minor gold and silver values in solid solution, as metals and as sulfosalts. Occurrence characteristics shift from disseminations through respective zones of microveinlets (crackle fillings), veinlets, veins and finally to individual structures on the periphery which may contain high-grade mineralization. Breccia pipes with attendant crackle zones are common. Expression of zoning is affected by exposure, structural and compositional homogeneity, and post ore faulting or instrusive activity. Vertical dimensions can reach 10,000 feet, with the upper reaches of the porphyry environment perhaps only at subvolcanic depths of a few thousand feet. Several lines of evidence suggest relatively shallow depths of formation and significant variations in water content in the porphyry environment. Shallow emplacement is consistent with the appearance of breccia pipes associated with ring and radical diking and with vertically telescoped zoning."

This rather complete description of porphyry coppers satisfies us as far as the geological controls are concerned. For the practical purposes of our metallurgical studies we can then state that a typical porphyry copper will be an oval or pipe-shaped deposit, roughly 3,500 x 6,000 feet in plane and up to a maximum of 10,000 feet in depth. It will contain an average of 140 million tons of ore, averaging 0.8 percent copper and 0.015 percent molybdenum and a variable quantity of pyrite. The primary zone of deposition will contain an average of only 0.45% copper, but the secondary enriched zone may vary from one to several percent in copper, depending on the situation. Some 70 percent of the ore is normally deposited in the intrusion itself, while 30 percent is in the surrounding country rock.

In Fig. 2.1 we presented a schematic cross section of a typical copper porphyry. It will be noted that the primary (hypogene) mineralization remained intact only in the lowest and most isolated zone of the orebody. In this case it contains an average copper content of about 1 percent and iron content of about 2 percent. Copper values are mostly in the form of chalcopyrite, although cases of hypogene bornite and particularly tennantite, tetrahedrite and others are known. Iron is mostly in the form of pyrite or pyrrhotite, and there are small quantities of molybdenite.

This primary mineralization in the upper parts of the orebody, close to the surface, was under intensive erosion and oxidation processes, which contributed to oxidation and decomposition of primary sulfides, mainly chalcopyrite and pyrite. Under the effects of atmospheric oxygen and water, pyrite decomposed into ferrous sulfate and sulfuric acid, while chalcopyrite formed copper oxides and sulfuric acid. The ferrous sulfate readily oxidizes into ferric sulfate and ferric hydroxide, and ferric sulfate hydrolyzes into ferric hydroxide and sulfuric acid.

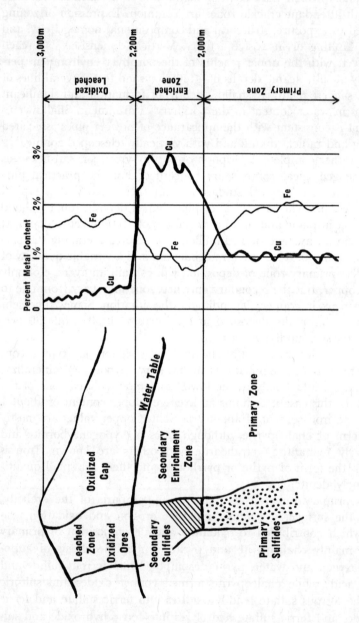

Fig. 2.1 A schematic cross section of a typical copper porphyry.

In this way through oxidation of the upper part of the original porphyry orebody, an oxidized zone is being formed with very favorable conditions for leaching and transportation to the greater depth of metallic values: copper oxides will readily dissolve into cupric sulfate which is then transported to a greater depth below the leached cap and water table, which roughly follows the contours of the surface. Ferric sulfate is an active solvent for copper sulfides and can also contribute to the dissolution of chalcopyrite and transportation of cupric ion at the greater depths. But the iron content of the ore, basically will not change in the oxidized cap, as Fig. 2.1 shows, because most of it will hydrolyze and precipitate in the form of iron oxides. The cupric ion will be transported into the secondary zone below the water table. It will react with primary copper and iron sulfides, thus readily forming the secondary copper sulfides. Reaction with pyrite will produce chalcocite, replacing hypogene pyrite volume for volume. Reactions with chalcopyrite may produce chalcocite, covellite, and bornite. Since each of these secondary minerals is higher in copper content than primary minerals, the overall result of this transportation of copper in greater depths is the formation of the secondary enrichment or supergene zone.*
As shown in Fig. 2.1, while the original copper content of the leached zone dropped from 1 percent to less than 0.5 percent, the secondary enrichment zone increased its original copper content from 1 to more than 2 and sometimes 3 percent copper. Some non-sulfide copper minerals such as chrysocolla or malachite and azurite, along with many others, could form as a result of the reaction of cupric ion with siliceous or carbonaceous gangue, but generally the proportion of these secondary non-sulfides is very low in comparison with secondary sulfides.

It should be clearly understood that historically, mining of copper porphyries was started earlier than 1906 when Jackling introduced the first mill to treat low-grade ores. Prior to this exploitation of the core part of porphyry coppers, an extensive mining of high-grade peripheral veins was widely practiced in many districts which later became famous for their porphyry copper deposits. To name a few we can refer to 19th century mining in the Copper Queen and Morenci district in Arizona, and early mining at Chuquicamata and El Teniente in Chile. Then, after the historic achievement of Jackling, mining of the central core part of copper porphyries was started. It was underground and

---

* Climatic conditions, such as are found in polar regions, may, however, reduce or minimize these processes of oxidation and result in minimal or no secondary enrichment.

mostly concentrated on the secondary enrichment zone, which contained more than 2 percent copper. This was the origin of the famous operations at Ray, Inspiration, New Cornelia, Miami and others in Arizona, McGill in Nevada, Chino in New Mexico, Bingham in Utah, Andes and Braden in Chile.

This first generation of copper porphyry operations was instrumental in collecting sufficient cash to build a new and improved technology, which would permit commercial treatment of lower grade ores found in the oxidized cap and in the primary zone. This gave rise to a second generation of plants, which, depending on the case, used open pit or underground mining. Typical representatives of this generation are Bagdad, Castle Dome, Copper Queen and Copper Cities in Arizona, Consolidated Coppermines in Nevada, Potrerillos in Chile, and Kounrad in Kazakhstan.

Finally, after the Second World War and particularly in the late fifties and early sixties there appeared a new generation of modern operations which could handle practically any low-grade ore above 0.5 percent copper. These were San Manuel, Silver Bell, Morenci and Ray in Arizona, Yerington in Nevada, Almalyk in Uzbeckstan, Toquepala in Peru, Cananea in Mexico, Atlas in the Philippines, and Majdanpek in Yugoslavia.

At the present there is being developed a new generation of porphyries, where in many cases copper content is not sufficient to justify the whole operation, but where by-product recovery, notably in molybdenite, gold and silver, helps to straighten out the economics. These are the cases of Brenda, Lornex, Gibraltar and Island Copper in British Columbia; Biga, Toledo and Santo Thomas in the Philippines; Bougainville in New Guinea; Medet in Bulgaria and several others in the process of development.

Genesis of Copper Deposits

It would now be interesting to say a few words about the genesis of copper deposits in general and of the copper porphyries in particular. This is because of the appearance of some new theories which throw an interesting light on possible zones of concentration of porphyry coppers in the world.

According to Cox and his collaborators,[8] copper is introduced into the accessible part of the earth's crust from unknown deeper sources by igneous intrusion and by upward migrating fluids. The economic concentration of this metal may result directly from these processes and from secondary effects such as weathering, erosion, and sedimentation. The porphyry coppers, like most important vein and replacement deposits, are genetically associated with the intrusion of felsitic igneous rock.

Here is how Cox et al.[8] describe this process:

"A part of the copper is trapped in disseminated grains by the rapid crystallization of the magma, which in turn gives rise to the characteristic porphyritic texture of intrusion. Another part of the copper is mobilized by water and other volatiles escaping from the hot congealed rock mass and is deposited in fractures in the intrusion and its wall rocks. A third part may escape completely from the intrusion and from vein and replacement deposits in nearby reactive host rocks. Thus porphyry copper deposits commonly occupy the central part of large and small mining districts containing vein and replacement deposits of copper, lead, zinc, silver, gold, iron and manganese."

An interesting, although still speculative theory about the origin of copper porphyries was forwarded recently by Sillitoe.[9] In what he calls a "plate tectonic model," Sillitoe postulates that porphyry copper and porphyry molybdenum deposits are the result of sea-floor spreading and faulting of lithospheric plates at continental margins. This theory, which offers a model for space-time distribution of porphyry deposits, indicates that chemical and isotopic data confirm that many porphyry ore deposits are formed by partial melting of oceanic crustal rocks on underlying subduction zones at the elongate compressive junctures between lithospheric plates.

The existence of copper-rich areas in the oceanic crust is explained by heterogeneous distribution of metals in the low velocity zone of the upper mantle. This copper is transported at divergent plate junctures laterally by ocean rises and accompanying basic magmatism. According to Sillitoe, "provinces with a high concentration of porphyry copper deposits, such as southern Peru, northern Chile and the Southwest of the United States, may be interpreted as regions beneath which anomalously copper-rich oceanic crust was subducted at the time of porphyry copper emplacement. . . . Porphyry ore deposits seem to have formed during a series of relatively short, discrete pulses, perhaps correlatable to changes in the relative rates and directions of motion of lithospheric plates. The time intervals during which the formation of porphyry deposits took place are shown to be broadly coincident with periods of lithosphere plate convergence, and porphyry deposits may still be forming above currently active subduction zones."

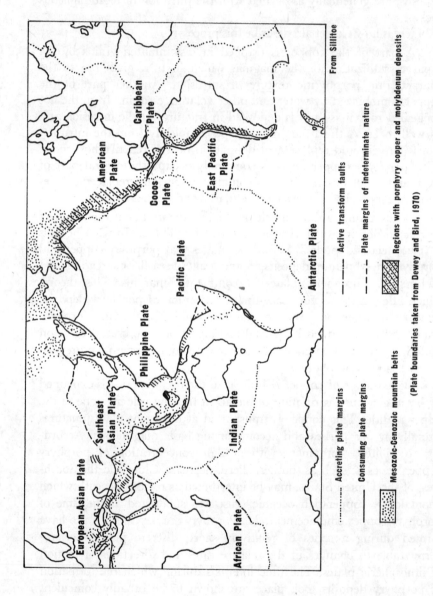

Fig. 2.2 Sillitoe's Plate Tectonic Model for the Origin of Porphyry Copper Deposits (from Economic Geology, Vol. 67, No. 2).

In Fig. 2.2, Sillitoe indicates boundaries of major lithospheric plates and indicates locations of porphyry copper orogenic belts. It proves that the majority of the world-known porphyry copper deposits are distributed around the Pacific and in the central portion of the Alpide Belt. The Pacific Belt extends from western Argentina, through Chile, Peru, Ecuador, Colombia, Panama, Mexico, the southwestern United States, Utah, Idaho, Washington, British Columbia, the Yukon and Alaska. Then it continues from Taiwan to the Philippines, Borneo, West Irian, New Guinea and the Solomon Islands. The Alpide Belt starts somewhere in South Banat in Rumania, crosses eastern Yugoslavia, central Bulgaria, Turkey and reappears in Armenia, Iran and western Pakistan. It can easily be observed how these porphyries are confined to orogenic belts (characterized by calc-alkaline magmatism) which apparently result from lithospheric oceanic plates underthrusting adjacent continental plates in what can be termed continental collision. The phenomenon is characterized by elongate contact delimiting the lithospheric plates.

Age of Copper Porphyries

In Fig. 2.3 a compilation of the age of porphyry copper deposits is made from the available bibliography.[9,10,11,12,13] As can be seen, the porphyry coppers' age ranges from less than 5 million to some 200 million years, i.e., spreads throughout the Cenozoic and Mesozoic, from the Pliocene to the Triassic.

The oldest copper porphyries of the Triassic and Jurassic ages, principally in British Columbia, were formed during the plate convergence when the North American Plate moved westward over the Pacific Plate.[8] The development of tensional structures in the continental crust due to the rise of the East Pacific Plate over the American later produced a new magmatic activity which gave rise to the Cretaceous and Tertiary deposits. These deposits, when developed in the thick continental crust, are generally associated with quartz monzonite,[7] while those formed in thin sectors of the crust are generally associated with quartz diorite. This is the case with all deposits which are older and developed on islands, such as Highland Valley, Copper Mountain, Bougainville and the Philippines deposits.

The majority of known porphyry coppers are of the Paleocene age. According to Lowell,[11] in the late Cretaceous time the main pulse of Laramide orogeny and intrusive activity and mineralization began. This continued for about 30 million years and was followed by a

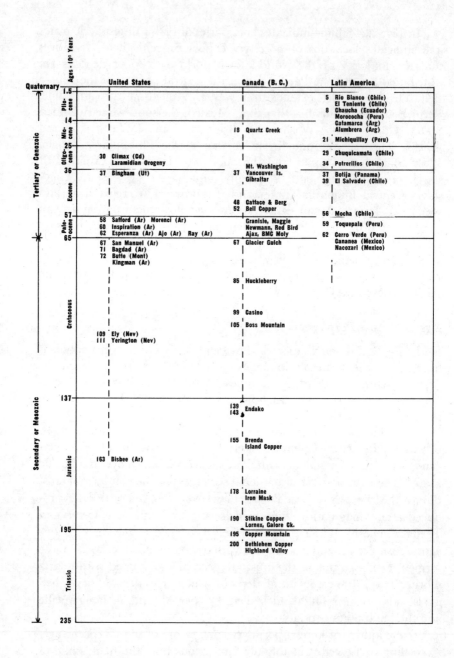

Fig. 2.3 *Age of porphyry copper deposits (from World Mining and other sources).*

second, weaker pulse in the middle Tertiary time. The same author shows a progressive decrease of age of porphyry orebodies from northwest to southwest among the Laramide deposits in Arizona. He concludes, then, that in very general terms this indicates that the locus of porphyry mineralization first shifted southward from British Columbia to Sonora and then shifted eastward for the deposition of the mid-Tertiary deposits.

Field et al.[12] report that the Pacific Northwest porphyries developed in a very complex region. Multiple orogenic episodes are recorded in variably deformed lithologies which in some regions are accompanied by regional metamorphism. Porphyry type deposits are generally found west of the Rocky Mountain Belt, and the rock units range from Precambrian through Cenozoic. The younger volcanoes which extend from northern California through southern Alaska are Pliocene and younger. The authors divide, then, the porphyry type mineralization into six age intervals starting from 210 million years and finishing at 15 million years. They conclude that at least two spatially and temporally distinct episodes of porphyry type plutonism and mineralization happened: one, between 210 and 190 m. years ago, which resulted in the deposition of such important copper-molybdenum orebodies as Bethlehem, Lornex and Ingerbelle-Copper Mountain; the other, younger event was between 45 and 70 million years ago and was confined to the Western Coast and was numerous in molybdenite-stockwork deposits, such as B. C. Molybdenum, Bell Copper and Granisle.

Hollister, in a recent paper,[13] stressed that although the porphyry copper deposits in the Andean orogen may have been formed from Permian time onward, the commercial deposits developed so far are not older than Cenozoic. Generalizations are, however, difficult because Andean deposits are not as well described as their counterparts in the Cordilleran orogen in North America. This refers also to many porphyries of the Alpide Belt. Sar Chesmeh in Iran is estimated at 12 to 15 million years, and the Bougainville orebody in New Guinea is only two million years old.

Wolfe[14] interprets the origin of the Philippine porphyry deposits through concepts of sea-floor spreading and plate tectonics. He considers that porphyries were formed as a result of the collision of crustal plates, subduction of the oceanic plate, emplacement of a serpentine ultrabasic diapir, the formation of a dioritic magma at depths of 300 to 500 kms, the rise of magma to a subcrustal or intra-crustal chamber, the eruption of andesitic volcanoes, the emplacement of diorite or differentiation of the magma into more acidic products, multiple in-

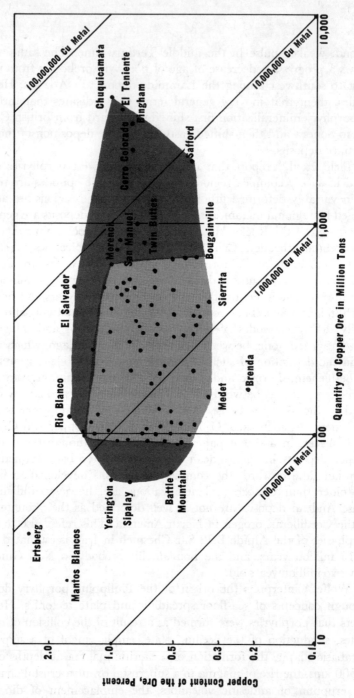

Fig. 2.4 Tonnage-grade relationship in copper porphyries.

trusion of hypabyssal rocks and hydrothermal alteration by fluids carrying metallic and other ions. He also believes that Philippine porphyries bear many similarities to Arizona porphyries, although some striking differences exist as well. Most of the Philippine porphyries are young, about 15 million years.

Economic Importance of Copper Porphyries

In order to systemize the tonnage-grade relationship among the copper porphyries, Fig. 2.4 was prepared. It indicates that the copper content of porphyry coppers ranges from a minimum of, say, 0.3 percent to an average maximum which generally does not exceed 1.7 percent. Cases like Brenda with only 0.2 percent copper content and that of Ertsberg with more than 2 percent copper assay are really exceptions. Most copper porphyries are within the 0.5% Cu to 1% Cu range, the fair average being at about 0.8% copper.

With respect to the ore tonnage, most of the copper porphyries are within a 50 million to 500 million ton range. There are smaller porphyries with only 20 million tons, like Mantos Blancos in Chile or Battle Mountain in Arizona, but then, in most cases, tonnage surpasses 100 million tons, and the copper content of these orebodies is close to or above 1,000,000 tons of metallic copper. The typical cases in our figure can be found in the lighter zone of the diagram.

Mines such as Bougainville, Morenci, San Manuel, Cerro Colorado, Safford and particularly Bingham and Chuquicamata are extraordinary cases in their huge potential.

It should be stressed that the figures considered for the diagram in Fig. 2.4 are average figures, and that in the same orebody very low-grade, average and high-grade ores can be found which make it susceptible to exploitation at different times and with different technologies. However, what makes a porphyry copper economically important is its average grade and potential by-product values which will justify a long-range investment.

Bibliography for Chapter Two

1. G. M. Schwartz: The nature of primary and secondary mineralization in porphyry coppers; Geology of the Porphyry Copper Desposits in Southwestern North America; Edited by S. R. Titley and C. L. Hicks; the University of Arizona Press, 1966.

2. A. B. Parson: The Porphyry Coppers in 1956; AIME Edition, New York, 1957.
3. S. R. Titley: Geology of the Porphyry Copper Deposits of Southwestern North America, p. IX, The University of Arizona Press, 1966.
4. H. E. McKinstry: Mining Geology, p. 650, Prentice-Hall Inc., Englewood Cliffs, New Jersey, 1948.
5. A. M. Bateman: Economic Mineral Deposits, p. 486, John Willey & Sons, New York, 1950.
6. J. D. Lowell: Copper Resources in 1970 — The 1970 Jackling Award Lecture; Mining Engineering, April 1970, pp. 67–73.
7. J. D. Lowell and J. M. Guilbert: Lateral and Vertical Alternation-Mineralization Zoning in Porphyry Ore Deposits; Economic Geology, Vol. 65, No. 4, June-July 1970, pp. 373–408.
8. Dennis P. Cox et al.: Copper — United States Mineral Resources, U.S. Geological Survey, Professional Paper 820, pp. 163–190.
9. Richard H. Sillitoe: A Plate Tectonic Model for the Origin of Porphyry Copper Deposits; Economic Geology, Vol. 67, No. 2, March-April 1972, pp. 184–197.
10. George O. Argall, Jr. and Robert J. M. Wyllie: Open Pit Copper Mines, World Mining, July 1973, pp. 34–37.
11. J. D. Lowell: Regional Characteristics of Southwestern North America Porphyry Copper Deposits; AIME Annual Meeting, Chicago, 1973, Preprint 73-S-12.
12. Cyrus W. Field et al.: Porphyry Copper-Molybdenum Deposits of the Pacific Northwest; AIME Annual Meeting, Chicago, 1973, Preprint 73-S-69.
13. V. F. Hollister: Regional Characteristics of Porphyry Copper Deposits of South America; AIME Annual Meeting, Chicago, 1973, Preprint 73-I-2.
14. John A. Wolfe: Tectonic Fingerprint in Philippine Porphyry Deposits; AIME Annual Meeting, Chicago, 1973, Preprint 73-S-37.

CHAPTER THREE

Economic Geography of Copper Porphyries

In Fig. 3.1 a polar view of the distribution of copper porphyries around the world is given. It clearly shows that the bulk of porphyry copper deposits is located in the zone which circles the Pacific Ocean, or, according to Sillitoe,[1] at the periphery of the Pacific Plate which expands in the direction of North and South America on one side and Southeast Asia and Australia on the other (see Fig. 2.2).

These deposits contain an estimated 247 million tons of copper discovered so far, as compared with the total of 260 million tons of copper estimated to be the total copper content of all porphyries of the world so far accounted for. The balance of 13 million tons of copper, or only 5 percent of the total, is being found in another large, Alpide zone which stretches from the Balkans through Asia Minor, Armenia and Iran, and reaches Pakistan. Porphyries are found all along this trace, and some additional orebodies have been discovered further to the north, in the Soviet Kazakhstan and Uzbeckstan. A porphyry copper deposit was recently discovered at Erdeinetin Obo in Mongolia,[2] and there have been reports of porphyry deposits in China. Thus it may be the case that the two belts, the Pacific and Alpide, cross somewhere, either in Taiwan or somewhere to the north. What is also clear is that the Alpide Belt is still very poorly explored, and thus holds a potential for substantially increasing copper reserves. The same is true for the Pacific Fire Belt.

On page 29 is an inventory of known copper reserves in different porphyry copper belts around the globe, which will be discussed in greater detail.

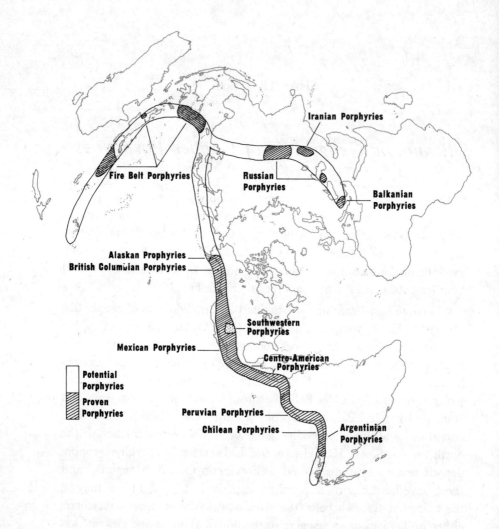

Fig. 3.1 Polar view of distribution of porphyry coppers.

The summary of presently known reserves, in terms of metallic copper, appears to be as follows:

Argentinian porphyries	— 2,000,000 tons
Chilean porphyries	— 86,000,000 tons
Peruvian porphyries	— 21,000,000 tons
Mexican & other Latin American	— 24,000,000 tons
United States porphyries	— 86,000,000 tons
British Columbian porphyries	— 15,000,000 tons
Pacific Fire Belt	— 11,600,000 tons
Alpide porphyries	— 13,000,000 tons
Total copper in porphyries	—260,000,000 tons

Latin America

The largest and richest porphyry copper deposits are located in Latin America, as Table 3.1 shows. They extend from western Argentina and southern Chile through northern Chile, all over the western coast of Peru, across into Ecuador, Colombia and Panama, and reappear in northern Mexico. The Mexican porphyries are genetically more related to the southwestern porphyries in Arizona. This Latin American copper belt may extend as far as Antarctica, because copper mineralization and porphyritic structures have been found on the Antarctic continent,[3] and it is claimed that the rock formations in this part of the world closely resemble those of the Chilean Andes.

It was previously believed that only the western slope of the Andes bore copper mineralization, thus leaving all porphyry copper deposits to Chile. This theory was, however, defeated in the sixties when United Nations geological teams conducted an exploration of the Andean slope in Argentina and found several porphyry copper deposits.[4] As Hollister stresses, most researchers now believe that South American porphyry deposits appear to be a function of internal failure near the leading edge of the continental plate, in which transcurrent and transverse strike-slip faults develop as a consequence of plate movement. The Argentinian porphyries in the southern part of the Andes appear to be of an older age than their Chilean counterpart: while the Paramillos Sur in Mendoza Province seems to be Triassic, La Disputada, El Rio Blanco and El Teniente across the Andes are only 4 to 5 million years old, i.e., from the late Pliocene. La Alumbrera and Mi Vida in Catamarca Province to the north are between 6 and 8 million years old.

The Argentinian copper porphyries are still poorly explored because of insufficient commercial interest. The best known is the Pachon orebody in the province of San Juan, which belongs to the St. Joe Co. So far about 170 million tons of ore containing over one million tons of metallic copper have been discovered. The grade of Argentinian porphyries seems to be lower than those of Chile, and although they also contain molybdenite as an accompanying mineralization, their economic significance is still in doubt. So far in three large deposits, about 400,000,000 tons of ore containing roughly 2,000,000 tons of copper have been discovered.

Chilean Porphyries

Chilean copper porphyries, along with those of the southwestern United States, can be considered the most important in the world. The ore reserves of these porphyries are estimated at over 8.5 billion tons, containing some 86 million tons of copper and over 2 million tons of molybdenum. The United States' reserves are somewhat smaller.

The Chilean copper belt starts at El Teniente, where one of the largest copper mines in the world is located. When in 1966 a "Chilenization" deal was accorded between Kennecott and the Chilean government, the following report of known copper ore reserves was officially given:[5]

Ore tonnage	% Cu	Copper content - tons
109,000,000	2.18	2,376,200
306,000,000	1.72	5,263,200
687,000,000	1.23	8,450,100
807,000,000	0.73	5,886,100
1,696,000,000	0.55	9,328,000
3,605,000,000	0.87	31,303,600

These are certainly the largest reserves ever discovered and officially reported in world history. While Chuquicamata or Bingham ore reserves are comparable or even possibly larger, no official information on them exists in comparable detail or extent.

Just less than 100 miles to the north of El Teniente is a medium-sized copper porphyry orebody presently worked by two companies: Andina and Disputada. Disputada, run by the Peñarroya French interests, is the older company and possesses a smaller part of the orebody with a richer ore. Officially, reserves are put at 50 million tons of 1.7 percent copper ore, but it is believed that more ore can be found with additional exploration, although of somewhat lower grade. The other

TABLE 3.1

DIRECTORY OF WORLD PORPHYRY COPPERS

NAME & LOCATION	PROPERTY OF	OPENED	Ore 10⁶t.	Cu 10³t.	Mo tons
Latin America					
1. Chuquicamata, Chile	CODELCO	1915	>2,500	>30,000	800,000
2. El Teniente, Chile	CODELCO	1906	3,500	31,500	1,050,000
3. El Abra, Chile	CODELCO	**	1,500	12,000	—
4. El Salvador, Chile	CODELCO	1960	300	5,500	120,000
5. Las Pelambres, Chile*	CODELCO	*	430	3,350	130,000
6. Rio Blanco, Chile	CODELCO	1970	120	1,900	18,000
7. Disputada, Chile	Penarroya	1962	100	1,400	14,000
8. Mantos Blancos, Chile	Hochshild	1961	20	320	—
TOTAL CHILE			8,700	86,000	2,150,000
9. Michiquillay, Peru*	MINEROPERU	*	575	4,000	150,000
10. Cuajone, Peru**	South.Peru	**	500	5,500	150,000
11. Toquepala, Peru	South.Peru	1960	400	4,000	80,000
12. Cerro Verde, Peru**	MINEROPERU	**	250	2,750	—
13. Morococha, Peru*	MINEROPERU	*	360	2,700	72,000
14. Quellaveco, Peru*	MINEROPERU	*	200	2,200	60,000
TOTAL PERU			2,300	21,000	520,000
15. Pachon, Argentina*	St. Joe	*	170	1,100	—
16. Paramillos, Argentina*		*	105	400	21,000
17. La Alumbrera, Argentina*		*	100	400	40,000
18. Chaucha, Ecuador*	Japanese Cons.	*	100	700	30,000
19. Antioquia, Colombia*	Colombia-USGS	*	625	6,250	—
20. Rio Vive-Takama, P. Rico	AMAX-Kennecott	**	240	1,750	—
21. Cerro Colorado, Panama	Canadian Javelin	**	>500	4,000	50,000
22. Pataquilla, Panama*	Japanese Cons.	*	300	1,800	—
23. La Caridad, Mexico**	Mexicana-ASARCO	**	600	4,500	100,000
24. Cananea, Mexico	Mexicana-Anaconda	1963	>500	4,000	—
25. La Verde, Mexico*	Mexicana	*	100	700	—
TOTAL OTHERS			3,400	26,000	250,000
TOTAL LATIN AMERICA			14,500	133,000	3,000,000

* In exploration stage   ** In development stage   *** Rough estimates

Fig. 3.2 South American Copper Porphyries.

half of the orebody was acquired by Cerro Corporation, which in 1970 inaugurated its 10,000 tpd. operation at Rio Blanco. This part of the orebody contains 120 million tons of ore assaying 1.58% Cu and some molybdenite values, which will also be recovered. The relationship between the two orebodies became evident only at a later date.

It is believed today when the present 1% Cu cut-off grade is lowered and some additional exploration is done, that the whole complex may grow into a substantially bigger property with several hundred million tons in ore reserves.

Farther to the north, in the province of Atacama and near the city of Copiapo are two porphyries: Potrerillos and El Salvador. Potrerillos is an old Anaconda property which was worked from 1927 to 1960, until its complete exhaustion. Then El Salvador was put into operation. Discovered during the Second World War, El Salvador was explored only in the fifties, when the eventual shut down of Potrerillos became evident. The original evaluation of 350,000,000 tons of 1.6 percent copper reserves still stands because additional exploration barely compensated for ore produced since the inauguration of the mine. The ore contains molybdenite and is mined underground.

In northern Chile, in the provinces of Antofagasta and Tarapaca lies what probably will prove to be the largest part of Chilean copper reserves. Apart from a relatively small porphyry at Mantos Blancos which, however, proved to be considerably larger than the originally claimed reserves of 10 million tons of ore, and originating in the huge orebody of Chuquicamata begins the so-called Fisher fault. This fault stretches from Chuquicamata farther in the north for at least 100 miles and comprises today about half a dozen recognized prophyry coppers and probably as many potential ones. When fully prospected and explored this porphyry copper belt will eventually prove to lock in several billion tons of ore with several dozen million tons of copper.

The giant Chuquicamata orebody, 2,500 x 10,000 ft. in horizontal dimensions and so far dug to the depth of 1,250 ft. contains at least 1.5 billion tons of 1.3% Cu proven reserves. Almost all of the mineralization is in the igneous host rock, the average grade being 1.7 percent copper and rather high 0.04 percent molybdenum. The hypogene ore assays 1 percent copper or somewhat higher and, essentially, is still not reached. The mine has already produced more than 13,000,000 tons

Fig. 3.3 Chuquicamata Metallurgical Complex. Mine in the background and mine dumps in the foreground. Exotica mine partially seen in the lower right corner, below mine dumps.

of copper and according to some semi-official evaluations, based on the geology and geometry of the orebody, has at least two to three times as much copper to be mined. This is in marked contrast to official reports which not so far back claimed that Chuquicamata had only 800,000 tons in copper reserves!

The exploratory drilling to the depth of 1,000 ft. below the bottom of the present pit proved a 1.5 to 2 percent copper ore, still in the secondary enrichment zone. Additional drilling to a further 1,000 ft. depth, i.e., some 3,300 ft. below the surface, indicates an ore in the 1.3 to 1.9 percent copper range. In fact, even at this depth many samples contained 2.1 and 2.3 percent copper.

As expected, with depth, mineralogical composition of the ore changes, because eventually the secondary enrichment fades out into the primary zone. This will decrease the proportion of chalcocite and markedly increase the proportion of pyrite and chalcopyrite, which will bear considerable influence on smelting characteristics of Chuquicamata concentrates in the future. According to a study, the following trend is likely to occur in the composition of Chuquicamata concentrates, expressed in percents.

Time period	Chalcocite	Covellite	Chalcopyrite	Pyrite
Actual	41	12	12	35
Next 5 years	29	4	3	64
5 - 20 years	22	8	18	52
after 20 years	10	14	35	41

Farther to the north from Chuquicamata extends the Fisher fault. Although first Andean intrusions are Jurassic, the rock age of this fault is estimated between late Eocene and early Oligocene.

Only 3 miles north of Chuqui lies the Pampa Norte orebody, which is a separate orebody but certainly genetically related to Chuquicamata. The orebody is buried under 250 to 350 ft. of alluvial overburden and is presently explored by drilling. It covers an approximate area of 10,000 x 3,000 ft. The oxide cap is 300 ft. thick and consists of 0.8–0.9% Cu ore. The secondary enrichment zone is very poorly developed and contains no more than 1 to 1.2% Cu, while the primary zone consists of chalcopyrite and bornite and assays an average of 0.8% Cu. There is a relatively small molybdenite mineralization.

Following the Fisher fault, farther to the north is the Paqui Range, consisting of intrusive rock similar to Chuquicamata, and presently under exploration. Then, 30 miles to the north from Chuqui is the large El Abra orebody. According to Jozsef Ambrus, Chuquicamata's chief geologist, so far about 500,000 m^3 of ore, or roughly 1.3 billion tons were indicated at El Abra. The drilled area covers some 300 million tons of ore containing 0.8% Cu and 0.01-0.02% Mo. This area is now being expanded with additional drilling into a 700 million m^3 reserve, which will be the core of a probable 1.5 billion ton orebody, when all outcrops are taken into consideration.

The oxidized cap outcrops right to the surface. It is between 250 and 450 ft. deep and consists of chrysocolla, atacamite and pseudo-malachite. It is estimated to be about 100 million tons of 0.4 to 1.0% Cu ore. The secondary enrichment is at least 1,500 ft. thick and not sufficiently explored. The primary zone consists of chalcopyrite and bornite and assays 0.7 to 0.8% Cu.

The orebody is also characterized by high-grade pockets such as Ojo de Gallo with 4% Cu ore, which is presently used for flux at Chuqui. Then there are exotic orebody manifestations, similar to Exotica at Chuquicamata.

The Fisher fault then continues farther to the north and contains Quebrada Blanca, a high-grade porphyry copper orebody, some 75 miles to the north of Chuqui; Cerro Colorado porphyry, formerly a Canadian property; Mocha — a well recognized copper porphyry belonging to the Chilean government and Copaquire, another porphyry, containing copper and molybdenum and very similar to Mocha.

This copperbelt also contains several exotic mines, such as Exotica, Sagasca and Huinquinlipa, which are not porphyries themselves, but which derive from some nearby porphyries. These orebodies also have high metallurgical values and as in the cases of Exotica and Sagasca are being exploited.

Attention was recently called to a new copper porphyry in central Chile at Pelambres, in the province of Coquimbo. This orebody was explored by a United Nations team which proved some 430 million tons of ore reserves containing from 0.5 to 0.9 percent copper and from 0.02 to 0.05 percent molybdenum, the overall averages being 0.78% Cu and 0.033% Mo.

The most extraordinary fact about Chilean copper porphyries is the high proportion of properties operated in comparison to the orebodies known. As Table 3.1 shows, out of 8 major porphyries known for a reasonable length of time, 6 are being exploited. This is a situation quite unusual for Latin America and is principally due to the high-grade of Chilean ore and a satisfactory investment climate, which prevailed before.

It is not difficult to foresee that with renewed confidence in the Chilean government and with very promising perspectives in the world copper market, new investment will be attracted to take advantage of fabulous Chilean copper resources in years to come. This area may eventually become one of the largest copper and molybdenum producing regions of the world.

Peruvian Porphyries

Another important group of copper porphyry deposits is found in Peru. It stretches from the southern Chilean border of Peru and goes up to the Ecuadorian border. So far about a dozen significant copper porphyries have been reported in Peruvian territory, but there is great potential for discovering at least as many more. Data on the most important Peruvian copper porphyries are reported in Table 3.1, and their locations are given in Fig. 3.2. So far some 2.3 billion tons containing 21 million tons of metallic copper and 520,000 tons of molybdenum have been discovered in these orebodies.

The main part of the Peruvian belt is located in southern Peru where Toquepala, Cuajone, Quellaveco and Cerro Verde orebodies are found. The only plant now in operation is at Toquepala, by Southern Peru, an American concern of four major United States companies: American Smelting and Refining Company, Cerro Corporation, Phelps Dodge and Newmont Mining. The Cerro Verde property which was confiscated from Anaconda is now being developed at a cost of $88 million by Minero Peru into a leaching operation which should be able to produce about 30,000 tpy. of copper by 1976. The other property, Cuajone, owned by Southern Peru, has recently finished all financial arrangements which will require at least $355 million to secure a 130,000 tpy. copper output by 1976. At Cuajone some 160 million tons of overburden should be stripped before the open pit mining can be started. A 30,000 tpd. concentrator will be constructed at the site. ASARCO holds the majority interest of 51.5 percent of Southern Peru Company.

The Cerro Verde property is being developed with Belgian credit, which along with Cerro Verde will be instrumental in the development of other orebodies, notably Ferrobamba, Chalcobamba, Antamina and Tintaya. Cerro Verde is only 15 miles from Arequipa and some 50 miles from the port of Matarani. In the first stage, the oxide cap will be removed and processed for copper recovery. After the sulfides, which account for about 90 percent of the 250 million tons of ore reserves, are exposed, flotation treatment for ores will be adopted, which probably will require another $100 million investment.

The Michiquillay porphyry deposit in northern Peru, which is considered to be the largest of all Peruvian copper porphyries and which was initially a concession of ASARCO, was turned over to Minero Peru because the $400 million investment necessary for its exploitation could not be subscribed by ASARCO. It has now received the attention of Japanese interests, consisting of nine Japanese smelters backed by Japan's Mining Association. The exploitation of this orebody is visualized at 130,000 of metallic copper per year, but the goal originally set of entering production by 1976 will obviously be postponed.

Other Latin American Porphyries

The search for copper porphyries in regions densely covered by vegetation in tropical and sub-tropical areas has recently brought very substantial discoveries. A medium-sized copper porphyry was discovered along the Chaucha fault in southern Ecuador. It is a typical continental expression of the oceanic structure which strikes in an east-west direction cutting across the northerly trending line of late Tertiary volcanoes.[8] So far some 100 million tons of 0.7 percent copper ore have been located, and the orebody is being explored by a Japanese consortium.

The other interesting find was made in northern Colombia, near Antioquia, by a U. S. Geological Survey and a Colombian exploration group. So far some 625 million tons of one percent copper ore have been reported.

A much larger strike was made in Panama by Canadian Javelin, which successfully developed ore reserves of up to 2.27 billion tons of 0.8 percent copper ore at the last counting.[9] Javelin estimates that its Cerro Colorado property contains ore reserves in excess of 3 billion tons with gold, silver and molybdenum values in addition to copper content. It plans a 176,000 tpd. milling operation, plus smelting and refining facilities at a total cost of investment of more than $0.5 billion An equivalent of 400,000 tpy. of copper will be produced in the form of concentrates, of which 50 percent will be shipped as concentrates, 200,000 tons smelted as blister and 100,000 tpy. locally refined. This may become one of the largest copper producers in the world along with Utah Copper, Chuquicamata and Udokan.[10]

Two Caribbean properties at Puerto Rico, somewhat away from the principal Cordilleran and Andean belt, have recently called attention to themselves due to ecological problems.[11] These are Rio Vive and Takama properties, accounting for some 240 million tons of 0.73 percent copper ore, owned by AMAX and Kennecott. These two companies spent some $10 million in acquiring land and for exploration of the orebodies, but were confronted with a strong ecological movement which resisted initiation of operations on this small island when they applied for mining leases. The Rio Vive orebody contains 104 million tons of 0.82% copper ore, and the Takama property has 139 million tons of 0.64% Cu ore, plus some gold and silver values. When and if in operation, the companies will produce some 50,000 tpy. of copper and 150,000 tpy. of sulfuric acid.

Mexican Deposits

For many years Anaconda's property at Cananea, Sonora, was the only operating porphyry copper in Mexico. The situation was largely due to the doubtful investment situation in that country and the relatively modest national exploration effort rather than to the unquestionable geological potential of this mineral-rich country. Things started to move in the early sixties with the rediscovery of the La Caridad orebody in 1960 by a joint Mexican-United Nations exploration team.[12] This mineral area known for its copper occurrences at present, was inactive for 70 years because of the different approach and exploration philosophies which predominated during all this time. When the need for copper and market conditions improved, active exploration was warranted.

La Caridad is located near Nacozari in the state of Sonora approximately 75 miles from the U. S. border, and belongs probably to the southern part of a well recognized copper belt which stretches from this region across to Arizona and then up to Bingham Canyon in Utah. The belt incorporates a cluster of porphyry coppers in Arizona and New Mexico. La Caridad is estimated to have 600 million tons of 0.75 percent copper ore. There is little doubt that this region may show additional copper porphyries.

United States Porphyries

Further to the north of the Mexican border lies a cluster of southwestern American copper porphyries, in the states of Arizona, New Mexico and Nevada. This can be considered the most important concentration of copper porphyries in the world, and an almost complete list of them and their geographical locations are given respectively in Table 3.2 and Fig. 3.4. Only the Chilean porphyries exceed in copper reserves those of the American Southwest, but then while the Chilean copperbelt spreads over more than 1,000 miles, southwest American coppers are found within a perimeter of 400 miles.

The North American copper porphyries are also the best studied and described orebodies. A few years ago an excellent summary of information available up to the mid-sixties was published by the University of Arizona in its "Geology of the Porphyry Copper Deposits of Southwestern North America." [12] Then, of course, there is the more updated study of Lowell and Guilbert published recently.[13]

As Table 3.2 shows, the largest known copper porphyry deposit in the United States should still be considered the Kennecott Company's Bingham Canyon Mine in the state of Utah. Although the official ore reserves are not available, from recent expansion of operations it is obvious that this large copper producer has at least a few decades to go, and at the present rate of 35 million tons of ore per year and 260,000 tpy. copper output, this means that the orebody is in the plus one billion ton range as far as the mineralized area is concerned, and probably as much as 10 million tons of metallic copper as far as production will go.*

The other two large mineralized areas on which official reports exist are Safford and San Manuel, both in Arizona. Safford is a very large and very low-grade region whose reserves are close to two billion tons, but assaying only 0.4 percent copper. This low-grade is still a limiting factor in future production plans. But the other, San Manuel deposit belonging to Magma Copper has been in production since 1956, and in its time was some kind of technological breakthrough for low-grade copper ores assaying only 0.6 percent copper. The San

---

* The only official statement about Kennecott Copper Corporation ore reserves is that in 1971 the operating properties (in Utah, Ray, Chino and McGill) had 3 billion tons in proven ore reserves which contained 18,000,000 tons of recoverable copper.[19]

TABLE 3.2

DIRECTORY OF WORLD PORPHYRY COPPERS (CONTINUED)

NAME & LOCATION	PROPERTY OF	OPENED	Ore 10⁶t.	ESTIMATED RESERVES Cu 10³t.	Mo tons
NORTH AMERICA					
26. Bingham, Utah	Kennecott	1906	>1,000***	10,000,000	750,000
27. San Manuel, Arizona	Magma Copper	1956	1,000	7,500,000	150,000
28. Morenci, Arizona	Phelps Dodge	1942	>500***	4,400,000	35,000
29. Butte, Montana	Anaconda	1964	>500***	4,000,000	—
30. Twin Buttes, Arizona	Anaconda-AMAX	1970	800	5,900,000	135,000
31. Sierrita, Arizona	Duval (Pennzoil)	1970	414	1,500,000	150,000
32. Pima, Arizona	Cyprus Mines	1957	200	1,100,000	26,000
33. New Cornelia, Ariz.	Phelps Dodge	1917	<500	3,750,000	—
34. Ray, Arizona	Kennecott	1955	<500***	4,000,000	50,000
35. Chino, New Mexico	Kennecott	1912	<500***	4,500,000	40,000
36. McGill, Nevada	Kennecott	1908	<500***	4,600,000	30,000
37. Tyron, New Mexico	Phelps Dodge	1969	<500	4,000,000	—
38. Inspiration, Arizona	Inspiration	1915	>500***	3,500,000	100,000
39. Mission, Arizona	ASARCO	1961	<100	500,000	—
40. Copper Queen, Arizona	Phelps Dodge	1885	<100	500,000	40,000
41. Mineral Park, Arizona	Duval (Pennzoil)	1964	<100	500,000	28,000
42. Esperanza, Arizona	Duval (Pennzoil)	1958	<100	600,000	5,000
43. Miami, Arizona	Cities Service	1954	<100	450,000	—
44. Silver Bell, Arizona	ASARCO	1954	265	2,120,000	50,000
45. Bagdad, Arizona	Cyprus Mines	1940			
46. Christmas, Arizona	Inspiration	1962	65	325,000	—
47. Battle Mountain, Ariz.	Duval (Pennzoil)	1967	55	500,000	—
48. Yerington, Nevada	Anaconda	1953	448	3,200,000	—
49. Lakeshore, Arizona	Hecla	**	350	1,400,000	35,000
50. Pinto Valley, Ariz.	Cities Service	**	2,000	8,000,000	—
51. Safford, Arizona	Kennecott	*	500	2,500,000	—
52. Florence, Arizona	Continental Oil	*	350	2,700,000	—
53. Metcafe, Arizona	Phelps Dodge	**	363	2,360,000	—
54. Helvetia, Arizona	Banner-AMAX	*	95	700,000	—
55. Palo Verde, Arizona	Banner-AMAX	*			
TOTAL UNITED STATES			13,000	86,000,000	1,700,000

* In exploration stage   ** In development stage   *** Rough estimates

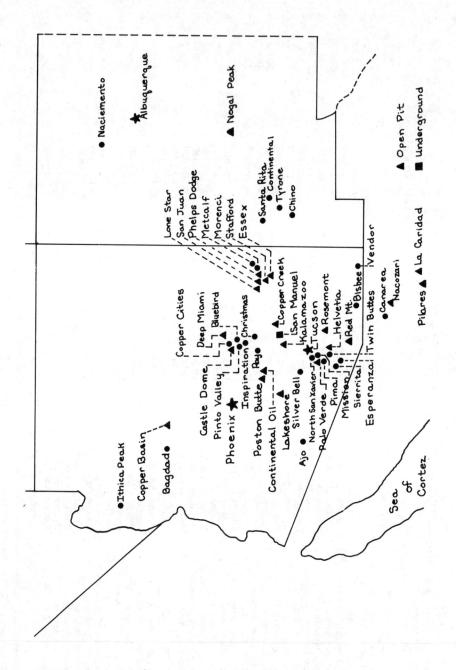

Fig. 3.4 Southwestern Copper Porphyries (reproduced from World Mining).

Manuel copper reserves were greatly expanded by the purchase of the Kalamazoo property (for $27 million), which constitutes a fault displacement segment of the San Manuel orebody. This acquisition triggered a new expansion of operations to 62,500 tpd. milling capacity and an annual output of copper which will reach 180,000 tons in 1974 at a cost of $250 million. Thus San Manuel will undisputably become the largest copper producer in Arizona.

The Morenci property of Phelps Dodge in Arizona and Butte of Anaconda in Montana still continue to be very important North American deposits although their copper grades are considerably lower. These old mining districts have passed through all phases of modern mining and milling and still vigorously produce copper from constantly discovered new reserves. At Metcafe, across the canyon and a few miles north of the Morenci pit, Phelps Dodge is developing a new mine[14] where another 350 million tons of ore have been found. Through 1972 some 46 million tons of overburden have been removed, and a 30,000 tpd. mining production will soon be on stream.

The early seventies were characterized by new copper ventures in Arizona, which reflect the American decision to rely more on its own production of the red metal in view of the changing world situation in the supply of this metal. The most important producers to inaugurate large new mills are Duval, with its extremely lean ore at Sierrita, and Anaconda at Twin Buttes, both in Arizona. Sierrita became economically feasible because of her large size and important molybdenum by-product recovery. Twin Buttes is now a joint venture with AMAX (after it acquired Banner) known under the name of Anamax. Both operations are based on very large ore reserves.

Among the properties now in active preparation for production are Pinto Valley, a large 40,000 tpd. milling operation owned by Cities Services, Lakeshore, a 9,000 tpd. Hecla operation, and the already mentioned Metcafe property of Phelps Dodge — all in Arizona.

The other area of great importance is the Pima-Mission-San Javier complex belonging to Cyprus, ASARCO and AMAX (former Banner interests). This alluvial-covered area contains probably as much as one billion tons of 0.5 percent copper ore, plus some molybdenite values. Portions of it have already been actively mined (Pima, Mission), with the exception of San Javier which has undergone only superficial stripping for oxide values.

The North American porphyry coppers extend from Arizona and Nevada farther to the north and northwest, into Utah, Idaho and Washington and then to British Columbia in Canada and into Alaska.

Fig. 3.5 Pacific Northwest Copper Porphyries.

A survey of the Pacific Northwest was presented recently.[15] Only Canadian properties in British Columbia undergo active exploitation, while the American properties in Idaho, Washington and Alaska are still economically inactive.

British Columbian Deposits

British Columbia has about a dozen recognized copper porphyries of which six are in operation and another one in the development stage. Lowell believes that there is great potential for another 100 copper porphyries in this mineral-rich country.[16]

As mentioned before, British Columbian porphyries are part of the Northwest Pacific metallogenic province, occupying the Western Cordillera mobile tectonic belt, which is about 2,100 miles long and 350 to 500 miles wide.[15] It stretches from California and Nevada in the south, through Idaho and Montana, Oregon and Washington, British Columbia and the southwestern Yukon Territory and southern Alaska, as shown in Fig. 3.5. Mines in active production are found only in British Columbia and almost exclusively in its southern and central part. The most active industrial zone is at Highland Valley, which some geologists claim may be potentially a mineral district comparable in importance to Bingham or to Butte. Here Bethlehem Copper with its several orebodies, Lornex and Valley Copper, is located.

Only a few years ago British Columbia was a minor copper producer. However, in 1971 her metal output was valued at some $93 million and probably doubled in 1972. Presently the installed copper and molybdenum output is in excess of $200 million. The principal copper producers are listed in Table 3.3.

The largest deposit discovered so far is the low-grade Valley Copper. Most of the ores are accompanied by commercial concentrations of molybdenite, whose recovery was decisive to the operation of most of them, particularly in the cases of Brenda, Lornex and Gibraltar. The British Columbian porphyry copper deposits are of lower than normal copper grade but their molybdenite grade is higher. A typical copper porphyry here will have about 0.6 percent copper and 0.03 percent molybdenite. They have a larger proportion of bornite, which in some cases surpasses the concentration of pyrite, and the secondary enrichment is less developed, probably because of climatic conditions. Recent glaciation may have been a factor.

Pacific Fire Belt

Another very promising area for porphyry copper, not yet properly explored but under intensive exploration, is certainly the Pacific Fire

TABLE 3.3
DIRECTORY OF WORLD PORPHYRY COPPERS (CONTINUED)

NAME & LOCATION	PROPERTY OF	OPENED	ESTIMATED RESERVES		
			Ore 10^6 t.	Cu 10^3 t.	Mo tons
NORTH AMERICA - Continued					
56. Gibraltar, B. C.	Placer	1973	358	1,325	37,000
57. Bethlehem, B. C.	Bethlehem	1962	376	1,850	65,000
58. Lornex, B. C.	Rio Algom-RTZ	1972	293	1,250	30,000
59. Island Copper, B. C.	Utah Constr.	1972	280	1,450	48,000
60. Brenda, B. C.	Brenda Mines	1970	165	300	54,000
61. Valley Copper, B. C.	Cominco-Beth.	**	750	3,600	—
62. Gaspe, Quebec	Noranda	1968	230	1,000	35,000
63. Liard Copper, B. C.		*	310	1,250	62,000
64. Maggie, B. C.		*	200	800	
65. Galore Creek, B. C.		*	100	1,000	
66. Granisle, B. C.		*	90	400	
67. Copper Mountain, B. C.		*	76	400	
68. Bell Copper, B. C.		*	46	230	
TOTAL CANADA			3,500	15,000	350,000
TOTAL NORTH AMERICA			16,500	101,000	2,000,000
TOTAL WESTERN HEMISPHERE			31,000	234,000	5,000,000

* In exploration stage   ** In development stage   *** Rough estimates

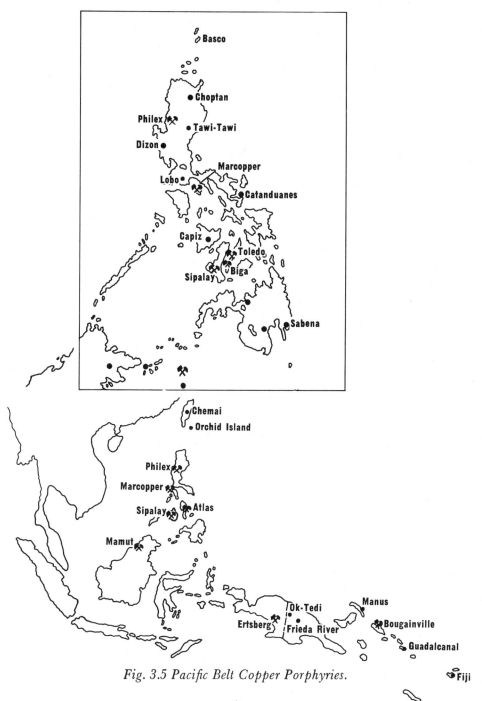

Fig. 3.5 Pacific Belt Copper Porphyries.

Belt. Attention to this area appeared during the Second World War when the mineral riches of this belt could be observed by the military personnel involved in the war. Now, as the investment situation deteriorates all around the world, this area has attracted particular attention.

The best explored and developed region in this area is certainly the Philippines. The largest porphyry deposit in the Philippines so far discovered is the Biga property belonging to Atlas (see Fig. 3.6). This 700 million ton orebody was put into operation in 1971 and now with Toledo, another Atlas porphyry copper, produces about 24 million tons of ore from which 66,000 tons of copper were recovered last year. The other working porphyries are Labo, a 200 million ton deposit belonging to Marcopper, Santo Thomas, a 125 million ton deposit of Philex, and the old Sipalay orebody of Marinduque started in 1957. A recent report [17] indicates that stripping of the Ino Capayang Bintakay property owned by Consolidated Mines has been started and that a 6,000 tpd. mill will be in operation in 1974. According to the same source Dizon and Taysan porphyries, respectively northwest and south of Manila, are being actively drilled. At the Tawi-Tawi property some 55 million tons of 0.5% Cu ore have been found and Lepanto Consolidated has been busy drilling properties on Mindanao (Surigao) and Luzon (Negros) Islands.

Another area of great attraction is New Guinea, the Solomon and Fiji Islands where several interesting prospects have been found in the last few years. Two of them, Ertsberg in the Indonesian part of New Guinea (West Irian) and Bougainville at Panguna at Bougainville Island, are already in operation, while the others are in the exploration stage. Kennecott located two porphyries just east of the West Irian border, Frieda River and Ok Tedi. Ok Tedi exceeds Frieda River both in grade and tonnage. Promising properties have also been found at Manus in Papua and New Guinea, at Guadalcanal in the Solomon Islands and Fiji Islands. The Mamut property in the Malasian part of Borneo is in the development stage by Japanese interests, and is due on stream by 1975. Some porphyries have also been detected in Indonesia, Taiwan and on Orchid Island. The Japanese and Russian sectors of the Fire Belt have as yet no reported porphyry copper deposits.

Alpide Porphyries

Most of this belt extends through the Communist Bloc countries, particularly Rumania, Yugoslavia, Bulgaria and the Soviet Union. Only a few porphyries are found in Iran and Afghanistan.

Rumania has several porphyry manifestations in Banat, but none of them is yet worked. The most prominent Yugoslavian porphyry, in operation since 1962, is at Majdanpek in eastern Serbia, near the Bulgarian border. Recently a very low-grade and large orebody was discovered at Krivelj, near Majdanpek. However, its commercial feasibility is still questionable. The best known Bulgarian porphyry copper is a 150 million ton deposit at Medet. It has been operated since 1965 and makes Bulgaria self-sufficient in copper. There are reports about porphyries in northern Greece, at Skouries, but so far no mining has been done in this area.

The next appearance of the Alpide Belt is in Soviet Armenia, in Transcaucasia. A cluster of several smaller porphyry copper deposits high in molybdenum content are distributed around Kadzharan, which is the largest of them all. Mining and milling operations have been reported also at Alaguez and Aigedzor, north and northeast of Erevan, the capital of Armenia. The cluster around Kadzharan includes Dasrakert, Kafan and Agarak, almost on the Iranian border.[18]

The only well known Iranian porphyry is the 300 million ton one percent copper Sar Cheshmeh, although other porphyries have been reported in the central and southwestern part of that country. One of them is Vazd, northwest of Sar Cheshmeh. A porphyry copper deposit has been reported in Afghanistan.

With respect to the Russian porphyry copper deposits the following can be said: there have been constant reports about a great exploration effort in Kazakhstan and Uzbeckstan, old provinces of Turkestan where the world's oldest copper was found. But although the Soviets claim about 200 promissory findings, only three of them, at Kounrad, Almalyk and Bozshchakul, are actually in operation. There is a mine at Saiak, east of the old Kounrad mine, which also supplies the Balkhash operation, but the other two large properties — at Dzhezkazgan and Udokan in Siberia — strictly speaking are not porphyry coppers but sedimentary coppers, similar to those in Zaire and Zambia. Thus, most Russian copper reserves, surprisingly, should be classified as sedimentary and vein type (principally in the Urals and North Russia and Siberia). The Russian copper porphyries, on the contrary, at this moment represent no more than, say, 15 to 20 percent of the known copper reserves. There is little doubt, however, that the porphyry copper reserves have great potential for growth as further areas of Central and East Asia belonging to the Soviet Union are explored. There is a great potentiality all along the border of China

TABLE 3.4

DIRECTORY OF WORLD PORPHYRY COPPERS (CONTINUED)

NAME & LOCATION	PROPERTY OF	OPENED	ESTIMATED RESERVES		
			Ore 10⁶t.	Cu 10⁶t.	Mo tons
ASIA & PACIFIC AREA					
69. Bougainville, N. Guinea	Bougainville	1972	760	3,600	—
70. Ertsberg, Indonesia	Freeport	1973	33	825	—
71. Mamut, Malasia	Japanese Cons.	**	120	840	—
72. Sar Cheshmeh, Iran	Iran	**	300	3,000	80,000
73. Biga, Philippines	Atlas	1971	700	3,500	—
74. Labo, Philippines	Marcopper	1969	200	1,100	—
75. Santo Thomas, Phil.	Philex	1958	125	600	—
76. Sipalay, Philippines	Marinduque	1957	60	500	24,000
77. Toledo, Philippines	Atlas	1954	100	600	—
TOTAL			2,300	14,600	104,000
*SOVIET UNION****					
78. Kounrad, Kazakhstan	USSR	1938	300	1,200	30,000
79. Bozshchakul, Kazakh.	USSR	1970	100	600	20,000
80. Kalmakyr, Uzbeckstan	USSR	1958	300	2,100	30,000
81. Kadzharan & Armenia	USSR	1950	200	2,600	100,000
TOTAL			900	6,500	180,000
EUROPE					
82. Majdanpek, Yugoslavia	Yugoslavia	1962	200	1,600	10,000
83. Krivelj, Yugoslavia	Yugoslavia	*	380	1,300	10,000
84. Medet, Bulgaria	Bulgaria	1965	150	540	15,000
TOTAL			750	3,500	50,000
TOTAL EASTERN HEMISPHERE			4,000	25,000	350,000
WORLD TOTAL			35,000	260,000	5,500,000

* In exploration stage ** In development stage *** Rough estimates

and Mongolia and also in Kamchatka, but at the moment, porphyry coppers are of secondary importance in the Soviet copper supply.

Bibliography for Chapter Three

1. Richard H. Sillitoe: A Plate Tectonic Model for the Origin of Porphyry Copper Deposits: Economic Geology, Vol. 67, No. 2, March-April 1972, pp. 184-197.
2. Intermet Bulletin, Vol. 2, No. 2, October 1972, p. 43.
3. George Mueller: Some notes on Mineralization in Antarctica; Antarctic Geology, SCAR Proceedings, 1963, VI Mineralogy.
4. V. F. Hollister: Regional Characteristics of Porphyry Copper Deposits of South America; Mining Engineering, August 1973, pp. 51–56.
5. Alexander Sutulov: Chilean Copper Production Expansion; Mining Magazine, London, July 1966, Vol. 115, No. 1, pp. 19–24.
6. V. D. Perry: History of El Salvador Development; Mining Engineering, April 1960, pp. 3–6.
William H. Swayne and Frank Trask: Geology of El Salvador, pp. 6–10.
7. Jorge Aviles: World's Largest Copper Reserve being developed in Chile; The Northern Miner, September 13, 1973.
8. P. J. Goossens: Metallogeny in Ecuadorian Andes; Economic Geology, Vol. 67, 1972, p. 462.
9. Javelin Contemplates Construction Phase of Cerro Colorado Project; Engineering and Mining Journal, August 1973, p. 26.
10. Cerro Colorado: The Continuing Story: Mining Journal, July 27, 1973, p. 65.
11. George Lutjen: The Curious Case of the Puerto Rican Copper; Engineering and Mining Journal, February 1971, pp. 74–84.
12. Spencer R. Tiley and Carol L. Hicks: Geology of the Porphyry Copper Deposits of Southwestern North America; The University of Arizona Press, Tucson, 1966.
13. J. D. Lowell and J. M. Guilbert: Lateral and Vertical Alteration-Mineralization Zoning in Porphyry Ore Deposits; Economic Geology, Vol. 65, No. 4, June-July 1970, pp. 373–408.
14. John V. Beall: Copper in the US — A Position Survey; Mining Engineering, April 1973, pp. 35–47.

15. Cyrus W. Field et al.: Porphyry Copper-Molybdenum Deposits of the Pacific Northwest; AIME Preprint 73-S-69, 1973.

16. J. D. Lowell: Copper Resources in 1970; Mining Engineering, April 1970, p. 67.

17. George Argall: Pacific Porphyries — More than You Think; World Mining, July 1972, p. 9.

18. Alexander Sutulov: Copper Production in Russia; Concepcion, Chile, 1967.

19. Frank Milliken: Summary Report of the Annual Meeting of stockholders — Kennecott Copper Corporation, May 4, 1971.

CHAPTER FOUR

Metal Production from Porphyry Coppers

Porphyry copper deposits are instrumental in the production of five metals: copper, molybdenum, rhenium, gold and silver. An estimated 43 percent of the world's new copper is currently being produced from porphyry deposits as shown in the following table:

Country	Total Copper Output	Output from Porphyries	Porphyries Percent
United States	1,780,000	1,600,000	90
Soviet Union	1,160,000	229,000	20
Chile	800,000	750,000	94
Canada	780,000	230,000	29
Peru	240,000	150,000	63
Others	3,340,000	541,000	16
Totals	8,100,000	3,500,000	43

The highest share of porphyry copper deposits is obviously in the Cordilleran and Andean areas, while the lowest is in Africa where sedimentary copper deposits are predominant.

With respect to the ratio of exploitation, i.e., production expressed as percent of reserves, the following can be observed: the industrially developed countries are consuming their deposits at a considerably higher rate than the underdeveloped world, which in the long-run will mean that the less developed countries will in time increase their importance in copper supplies. The situation can be appreciated from the following comparison:

	Reserves tons	Production tons	Production % / Reserves
Soviet Union	6,500,000	229,000	3.52
United States	86,000,000	1,600,000	1.86
Canada	15,000,000	230,000	1.53
Chile	86,000,000	750,000	0.87
Peru	21,000,000	150,000	0.71

So it may be safely concluded that much of the future copper supply will eventually come from Chile, Peru, Latin America and the Pacific Fire Belt. A rather similar situation can be observed with the sedimentary coppers of Zaire and Zambia.

Copper porphyries are also important suppliers of by-product molybdenum. It should be clearly understood that the bulk of the molybdenum supply comes from porphyry-type deposits also. The best known among them are Climax, Henderson and Questa in the United States and Boss Mountain, Endako and British Columbia Molybdenum in Canada. These mines, where molybdenite mineralization is predominant and where only traces of copper and other metals are found, account for approximately 78 percent of the world's known molybdenum reserves. The other 17 percent of the world's molybdenum reserves are found in porphyry coppers and only the balance of 5 percent are contact-metamorphic zones of tactite bodies, silicated limestone adjacent to intrusive granitic rocks, pegmatites, bedded deposits in sedimentary rocks, etc. A recent study[1] put the world's identified molybdenum resources at about 31.5 million tons, which means that approximately 30 million tons are in porphyry deposits. Of this total some 24.5 million tons of molybdenum are in deposits where molybdenite is the sole or the most important mineral, while another 5.5 million tons are in copper porphyries.

A similar situation is reflected in world molybdenum production. Of the 1970 total molybdenum output of 82,000 tons,[2] some 50,000 tons of molybdenum came from molybdenite porphyries, 30,000 tons from copper porphyries and less than 1,000 tons from other deposits. It is expected that in the future the share of molybdenum porphyries will greatly increase because of the huge production planned at the Henderson-Urad complex in Colorado, although by-product molybdenite production has also a high potential for growth. In fact, since 1970 the new operations at Sierrita, Twin Buttes and Pima in Arizona and at Lornex, Gibraltar, and Island Copper in British Columbia have added some 9,500 tons per year in additional molybdenum production potential.

Copper porphyries are also the sole source of the metal rhenium. As is known, this modern and relatively little known metal, which now enters into important applications in catalysis, environmental controls, alloys and space exploration, does not form individual mineral species but rather isomorphically replaces molybdenum in molybdenite. Thus, molybdenite is the only commercial source of rhenium. However, it happens that rhenium preferentially concentrates in molybdenite associated with copper minerals in copper porphyries. For example, while the concentration of rhenium in molybdenite from molybdenum porphyries barely amounts to 1 to 10 ppm, in copper porphyries molybdenite values bear from 100 to 2,000 ppm of rhenium. We thus submit that rhenium assays of molybdenite values in porphyry deposits may serve as an indicator for copper or molybdenum porphyry. For example, while the majority of Chilean and American copper porphyries contain molybdenite assaying between 300 and 600 ppm, molybdenite from Brenda Mines in British Columbia, where the ore is rich in molybdenum and poor in copper, contains only 80 ppm of rhenium, and Climax molybdenite contains less than 5 ppm of rhenium.

Finally, copper porphyries are a significant source for gold and silver. The gold and silver content of porphyry copper ores is not high, probably from 0.01 to 0.02 oz per ton for gold and from 0.05 to 0.1 oz per ton for silver. Nevertheless, even such small quantities of gold and silver are converted into important production by the sheer tonnage rate at which porphyry coppers are processed. Gold and silver are normally recovered in the refining stage of production and roughly 1 to 1.5 oz of gold and 3 to 10 oz of silver can be obtained per ton of copper. Naturally, some exceptions exist. The production figures on gold and silver recovery and copper porphyries are rather sparce, but some idea can be obtained from Table 4.1, which is the official account of the Kennecott Copper Corporation Metal Mining Division on metal production in the last decade.[3] Because of the size of this company, which gets all of its copper from some of the world's largest copper porphyries and, in fact, accounts for about 1/7th of the total output, this table is interesting for general analysis.

It will be noticed that typically one ton of ore mined will require about another 3 tons of material moved to the dumps. Then 1 ton of ore will typically produce from 15 to 16 lbs of copper. This, of course, is not true for many porphyries in British Columbia and Arizona, which

Fig. 4.1 Fingerprint of the Kennecott giant open pit at Bingham, Utah. (Photographed by Don Green, courtesy of Kennecott Research Center.)

give considerably lower yields, and for Chile and Peru in Latin America, where yields are higher. However, taking into consideration that 538 million tons of porphyry copper ores processed per year have the potential of producing 3.63 million tons of copper, tends to make this statement fair, because on a world-wide basis today a ton of porphyry copper ore yields 13.5 lbs of copper.

As far as molybdenum output is concerned, the Kennecott figures are somewhat distorted because a part of molybdenum production, at least from 1965 on, has been derived from the molybdenum mine in British Columbia. When U.S. molybdenum production alone is taken into consideration, one ton of ore yields 0.23 lbs of molybdenum as compared with 0.14 lbs Mo for the world average (including tonnage from all porphyry coppers, which do not produce molybdenum by-product). This gives us approximately 30 lbs of Mo per ton of copper for Kennecott, which has molybdenum by-product recovery installed in all of its plants, as compared with 21 lbs for the world-wide average. Only 28 out of 50 prophyry copper operations recover molybdenite by-product. They represent 65 percent of the ore tonnage and 66 percent of copper output. In other words, if corresponding corrections are made, a typical copper porphyry, which recovers molybdenum as a by-product, produces 0.22 lbs of Mo per ton of ore and 32 lbs of Mo per ton of copper — figures which are almost identical with the Kennecott averages.*

Kennecott is also the foremost producer of rhenium in the world. It has an installed capacity of about 4,000 lbs per year, one half of the world total,[4] and probably produced half of the reported world output of 5,900 lbs in 1970.[5]

The gold and silver recoveries at Kennecott average 0.75 oz of gold and 9.4 oz of silver per ton of copper. They are 1.67 oz of gold and 6 oz of silver per ton of Pacific copper, which is richer in noble metals (see Table 4.4). If we apply the lowest averages on all porphyry coppers, then those at 3.5 million tons of copper output could produce some 2.6 million oz of gold and 21 million oz of silver, which is respectively 5.5 and 7 percent of the 1970 production. The percentage is relatively small, but the value at the present prices exceeds $350 million, which is about $100 per ton of copper produced. This is two times

---

* A recent US Bureau of Mines study[43] indicates that at $0.50 per pound copper price 83 million tons of metallic copper could be recovered economically in the United States along with 1.8 million tons of molybdenum, 43 million oz of gold and 740 million oz of silver. This means that recovery of each ton of metallic copper will result in recovery of 44 lbs of by-product moly, 0.5 oz of gold and 9 oz of silver.

TABLE 4.1
PRODUCTION DATA OF METAL MINING DIVISION OF KENNECOTT COPPER CORPORATION
(operations include US and Canadian subsidiaries)

	Copper Ore mined, milled and treated 10^6 tons	Material Removed to dumps 10^6 tons	Copper Produced net tons	Grade of Ore Percent copper	Mo produces pounds	Au prod ounces	Ag prod ounces
1963	47.1	122.4	381,089	0.804	11,525,000	306,613	2,502,353
1964	44.3	111.0	368,184	0.821	11,315,000	261,086	2,077,592
1965	56.5	151.1	451,645	0.822	16,347,000	405,015	3,280,431
1966	57.9	132.6	454,044	0.792	15,577,000	387,727	4,763,348
1967	33.8	69.9	289,016	0.778	9,853,000	214,689	2,769,292
1968	47.2	120.2	378,215	0.747	17,023,000	290,594	3,229,258
1969	66.7	164.6	495,968	0.749	21,471,000	442,339	3,863,239
1970	68.6	179.8	518,888	0.773	23,124,000 (5)	404,141	4,338,730
1971	59.3	157.7	456,142	0.796	18,460,000 (1)	340,636	3,711,141
1972	58.5	164.0	460,576	0.787	15,041,000 (2)	350,080 (3)	4,335,074 (4)

Source: Kennecott Annual Report for 1972

(1) US Production 13,353,200 lbs. and B.C. Production 5,107,000 lbs.
(2) US Production 13,999,600 lbs. and B C. production 1,061,400 lbs.
(3) Virtually all copper by-product
(4) Of this 3.2 million or derived as copper by-product
(5) US Production 16,981,600 lbs. and B.C. production 6,141,300 lbs.

more than producers receive for 32 lbs of molybdenum contained in the by-product concentrate.

U. S. Production

In Table 4.2 a general summary of United States copper and molybdenum production is given according to the best information available by mid-1973. This information indicates that at the present, approximately 256 million tons of ore are processed annually to yield 1,700,000 tons of copper and some 45,000,000 lbs of molybdenum. The copper and molybdenum production capacity has recently been increased significantly by putting on stream the Sierrita and Twin Buttes deposit in Arizona, expansion of San Manuel and Pima, and the inauguration of Tyrone operations in New Mexico. Further expansion, which reflects an increased preoccupation by American industry for its raw materials supply and the general insecurity of foreign investment in the world, is expected very soon at Pinto Valley, Lakeshore and Miami in Arizona by 1974, and at Bagdad Copper, Metcalf and Twin Buttes in 1975, and several oher operations later in the seventies (see Table 4.7). This will contribute between 250,000 to 300,000 tons in annual new copper production.

Kennecott Copper Corporation which is the largest single producer of copper in the United States in 1970 reached a record output of 518,888 tons of copper, fully 30 percent of the U.S. 1.72 million tons of domestic mine production. The company's production data, shown in Table 4.1, demonstrate a steady effort of this company to expand its domestic production in view of the lost sources of supply in Chile. Its former Chilean subsidiary El Teniente had a production capacity of 200,000 tpy of copper and 3.8 million pounds of molybdenum, and by 1971–1972 should have reached a 280,000 tpy copper output, an effort which was crippled by the nationalization of this company by Allende's government and also by deficiencies in water supply and new smelter design.

As Table 4.1 shows, the domestic copper output of Kennecott gradually rose from 380,000 tons to 520,000 (with the exception of curtails for market or labor difficulties). The present capacities of her four divisions in Utah, Arizona, Nevada and New Mexico can be seen in Table 4.2.

The expansion of operations in several recent years includes: addition of the new Bonneville crushing and grinding facilities which increased the Arthur and Magma concentrating plants from 90,000 to

TABLE 4.2
PRODUCTION DATA ON OPERATING U.S. COPPER PORPHYRIES

Name & Location	Ore Production – tons		Heads		Annual Output – tons	
	Annual	Daily	% Cu	% Mo	Cu	Mo
1. Magna & Arthur, Utah	35,000,000	108,500	0.69	0.03	320,000*	6,500
2. San Manuel, Arizona	21,500,000	65,000	0.69	0.015	144,000	2,500
3. Butte, Montana	18,000,000	50,000	0.76	–	115,000	–
4. Morenci, Arizona	18,000,000	60,000	0.83	0.015	180,000*	Curtailed
5. Sierrita, Arizona	25,750,000	83,000	0.29	0.03	70,000	5,800
6. Pima, Arizona	18,000,000	54,000	0.56	0.017	82,500	1,000
7. Tyron, New Mexico	14,400,000	48,000	0.89	0.012	100,000	–
8. Twin Buttes, Arizona	10,700,000	32,000	0.6	0.03	75,000*	1,100
9. Ray, Arizona	10,500,000	25,500	0.98	0.015	90,000*	350
10. New Cornelia, Arizona	10,250,000	34,000	0.70	–	62,000	–
11. Yerington, Nevada	10,700,000	30,000	0.86	–	78,500*	–
12. Chino, New Mexico	7,250,000	22,000	0.90	0.008	73,500*	250
13. McGill, Nevada	7,700,000	21,500	0.85	0.016	48,000	250
14. Inspiration, Arizona	6,800,000	22,000	0.71	0.007	54,000*	220
15. Mission, Arizona	8,000,000	22,500	0.7	0.02	45,000	1,100
16. Mineral Park, Arizona	7,650,000	19,000	0.42	0.03	26,500*	1,750
17. Copper Queen, Arizona	5,350,000	16,000	0.83	–	48,000*	
18. Esperanza, Arizona	5,250,000	15,000	0.37	0.03	20,000	1,250
19. Miami, Arizona	4,500,000	14,000	0.50	0.005	22,300*	110
20. Silver Bell, Arizona	4,200,000	13,000	0.7	0.01	20,000*	350
21. Bagdad, Arizona	2,000,000	6,000	0.7	0.03	20,000	230
22. Christmas, Arizona	1,600,000	5,500	0.80	–	11,500	–
23. Battle Mount., Arizona	1,650,000	4,500	0.84	–	16,500*	–
TOTAL UNITED STATES	256,000,000	775,000	0.74	0.01	1,722,000	23,500

* Part of production obtained either by leaching ore or dumps
Contains 0.5% Cu Sulfide which is treated in a flotation plant (14,000 tyd) and
0.35% Cu which is recovered in a 16,000 tpd. leaching plant

108,500 tpd; construction of a new cementation plant which recovers copper from leached dumps and added some 50,000 to 60,000 tpy of copper production to the Utah Copper Division; construction of a leaching and electrowinning plant at Ray, which treats 10,000 tpd of oxide ore for recovery of 24,000 tpy of metallic copper; expansion and modernization of smelters in Utah, Chino and McGill.

These changes were consistent with general tendencies observed today in the United States: greater movement toward self-sufficiency; increased copper recovery from old dumps and by leaching tails; more attention to the by-product recovery, including sulfur from smelter gases; emphasis on low-cost open pit methods. Today 84 percent of the copper output of the United States comes from open pit mines, most of them porphyry coppers; and some 10 percent of the mine output, i.e., 172,000 tons in 1970, is recovered by dump and in-place leaching[6] and precipitation with iron. Another 70,000 tpy of copper is obtained by leaching ores, of which half is precipitated by iron and another half is electrowon. John Beall reports[7] that over 200,000 tpy of copper is recovered as precipitates at different copper porphyries which substantially improves the overall economic efficiency of porphyry copper deposits. The major precipitation plants and their capacities are as follows:

Utah Copper	50,000 tpy Cu
Chino Mines Division	20,000 "
Ray Mines Division	40,000 "
Butte, Montana	24,000 "
Yerington, Nevada	20,000 "
Inspiration Copper	23,000 "
Copper Queen-Lavander	17,200 "
Duval - 3 plants	14,000 "
San Xavier-ASARCO	11,000 "
Bagdad Copper	7,300 "
Miami-Copper Cities	8,400 "
Others	13,500 "
Total Capacity	248,400 tpy Cu

San Manuel of Magma Copper Company, a 100 percent owned subsidiary of Newmont Mining Corporation, which was put on stream in 1956 as a 30,000 tpy operation with an annual copper of 70,000

Fig. 4.2 General view of Kennecott's new cementation plant at Bingham, Utah. (Photograph by Don Green, courtesy of Kennecott Research Center.)

tpy, has now been expanded to more than two times its original capacity. After the acquisition of Kalamazoo, San Manuel copper ore reserves doubled to over 1 billion tons of ore averaging 0.72% Cu.[8] This acquisition justified expansion of operations from 40,000 tpd in 1968 to over 60,000 tpd today. The expansion of production was achieved by sinking two new shafts, expansion of the San Manuel concentrator, smelter and refinery. The whole expansion-modernization program in three years cost $250 million including a refinery with a 200,000 tpy capacity which also should take care of copper refining from another Magma subsidiary, Superior, which treats high-grade copper and gold ore. The new installations include copper finished products such as continuous cast 5/16 in. copper redraw rod, full plate cathode copper and sheared cathode plate.

Another large newcomer on the American copper scene is Duval, a wholly owned subsidiary of Pennzoil. Duval, previously known as Duval Sulphur and Potash, a subsidiary of United Gas, was acquired by Pennzoil United, a large Houston-based natural resources company, when United Gas and Pennzoil consolidated in 1968. Duval has operated since 1958 a 15,000 tpy copper mine at Esperanza and since 1964 another medium-sized copper producer at Mineral Park. In 1967, Duval initiated operations at a smaller property at Battle Mounain. Treating the ores which generally contain about 0.5 percent copper at costs below $3 per ton, Duval was able consistently to make revenues of $3.64 to $4.86 per ton[9] and this apparently stimulated its involvement in a new and large $200 million investment at Sierrita. Development of this project was partially financed by the U.S. government, according to a policy which stimulates national production by government loans which are later repaid by deliveries of refined copper to the national stockpile. Sierrita is a large, almost 0.5 billion ton low-grade copper porphyry. Its copper content is very low according to present standards -- only an average of 0.32% Cu — but it is complemented with substantial molybdenum mineralization amounting to 0.036% Mo. This, at the present expanded rate of production of 83,000 tpd makes possible an annual production of copper of 70,000 tpy and a molybdenum output close to 12 million pounds. Also some 500,000 oz of silver are recovered annually. The U.S. government will receive in the first five years of operation 109,000 tons of copper at $0.38 per pound.[10] So far the Sierrita operation has proven to be a success: it will easily repay $83 million in government loans, since it rapidly reached the designed capacity, and its 1971 gross revenues per ton of ore milled were $3.25 per ton, as compared with costs of $2.67.

Fig. 4.3 Early picture of Anaconda's Berkley open pit in Montana. (Courtesy of George O. Argall, Jr. of World Mining.)

Depreciation and depletion costs which should be added are $0.32 per ton of ore.[11] Sierrita also converted Duval into one of the most important molybdenum producers in the United States, after AMAX and Kennecott. Sierrita molybdenum concentrates contain a medium concentration of rhenium, of about 170 ppm, which is probably related to the low-grade copper mineralization. The Brenda molybdenite concentrates in Canada have only 80 ppm of rhenium and this is with ore assaying only 0.2% Cu.

Another very dynamic enterprise is the Pima Mining Company, a subsidiary of Cyprus Mines. It was discovered in 1950 by geophysical methods and was thought to be a small, high-grade orebody. This is the reason why in 1957 it was initiated as a small 3,500 tpd mill. Since that time, this area 20 miles southwest of Tucson, covered with a 200 ft. layer of aluvium, proved to be a large mineralized copper area with ore reserves in the 1 billion plus ton range. Pima owns some 200 million tons, assaying an average of 0.56% Cu plus some molybdenum values. For this reason the spectacular expansion of this small mill into one of the largest operations in the United States has taken place, following the schedule given here:

1957	3,500 tpd	1.74% Cu
1963	6,000 tpd	1.34% Cu
1966	18,000 tpd	0.73% Cu
1968	30,000 tpd	0.58% Cu
1970	40,000 tpd	0.54% Cu
1972	54,000 tpd	0.56% Cu

Since 1963, the Pima copper output has grown almost four times from 22,000 tpy to 82,500 tons in 1973. Pima also produces today about 1,000 tons of molybdenum per year.

Another large porphyry copper property in the same area is Twin Buttes, an 800 million ton orebody, jointly owned by Anaconda and AMAX, under the name of Anamax. This ore deposit is also covered by a layer of hard and tough conglomerate and some 400 to 600 feet of alluvial overburden which had to be removed before the operations could be started. This required the removal of 236 million tons of rock before the first pound of copper was recovered, a volume sufficient to construct the Aswan Dam in Egypt. The orebody contains an estimated 6 million tons of metallic copper and 700,000,000 lbs of molybdenum, a metal content very important for Anaconda as a major copper producer in the world after she lost her fabulous properties in Chile which

Fig. 4.4 ASARCO's newest copper leaching plant at San Xavier, Arizona. (Photograph by George O. Argall, Jr. of World Mining.)

at one time secured for her some 600,000 tpy of annual production, of which 2/3 was supplied by Chilean subsidiaries. In fact, the 1969 Anaconda copper output came from the following sources:[12]

Montana	Berkley Pit	75,104	tons
	Vein Mines	29,536	tons
Nevada	Yerington	51,232	tons
British Columbia		6,383	tons
Mexico	Cananea	36,521	tons
Chile	Chuquicamata	312,157	tons
	El Salvador	84,927	tons
	La Africana	2,181	tons
Total		598,241	tons

It is obvious that after nationalization of Chilean properties, during which the company lost 67 percent of its copper output and 75 percent of its profits, Anaconda had to fight for its life as a major copper producer. In the 1971–1972 period a drastic reorganization took place in view of a reported loss of $357 million out of total revenues of $946 million in 1971. Some $348.5 million of the loss was written off on its Chilean subsidiaries. A new plan was set up to develop the company's copper output to a 300,000 tpy level by investing some $200,000,000. This plan, due by 1975, contemplated: 1) the development of production levels at Montana and Nevada properties of about 200,000 tpy; 2) construction of a 30,000 tpd flotation mill at Twin Buttes, with an annual output of about 70,000 tpy; 3) the expansion of the same mill to 40,000 tpd with an additional copper output of about 25,000 tpy; 4) construction in cooperation with AMAX of a 10,000 tpd leaching operation for copper oxides at Twin Buttes, which should provide some 30,000 tpy of copper. The greater part of these programs is already fulfilled, and the last additions are well on the way. So it is almost sure that by 1975 Anaconda will surpass a 300,000 tpy annual output.[13-14] This is particularly important because in the meantime, Mexico has also "Mexicanized" 50% of its copper property at Cananea, although under negotiated and favorable conditions.

American Smelting and Refining Company (ASARCO) mines most of its copper in the United States in medium-sized operations. The largest one is at Mission, a 22,500 tpd mill producing an equivalent of 45,000 tpy of copper. Mission is adjacent to the Pima Mining Company with which it apparently shares the same orebody. Just

Fig. 4.5 Loading first ore at Mission mine of ASARCO. (Courtesy of George O. Argall of World Mining.)

north of Mission, ASARCO has an oxide copper pit at San Xavier, the ore from which is leached near the Mission concentrator. This adds another 11,000 tpy of copper. When the oxide cap is eliminated, the underlying sulfides will be processed at the Mission flotation plant.

The other ASARCO property is at Silver Bell, where a 7,500 tpd mill was constructed in 1952. Presently this operation has reached a 10,500 tpd capacity, and produces about 15,000 tpy of copper in concentrates and some 5,000 tpy of precipitates from dump leaching.

Near Casa Grande, ASARCO is developing a new mine at Sacaton. This is a 50 million ton 1% Cu orebody overlaid by 45 million tons of alluvium which must be removed before the ore is mined. A 9,000 tpd concentrator is being built at a cost of $35 million for the whole project. The property will be operated as an open pit mine until the end of this decade, and then should go underground. It will probably come into operation in 1974.

Cities Service has also been very active in the development of its copper properties. In addition to its plant in Tennessee, which is not a porphyry copper, Cities Service is expanding its operations in the Miami-Copper Cities area and breaking in a new large property at Pinto Valley. At Miami, one of the oldest mines started in 1911 is now being leached in situ, yielding about 6,000 tpy of copper in the form of precipitates. At nearby Copper Cities a 14,000 tpd flotation plant is processing a copper-molybdenum ore containing as little as 0.5% Cu and 0.005% Mo. Now, in the same area, a new 50 million ton deposit assaying about 2% Cu is being developed at a depth of about 3,000 feet. The ore shipment to Copper Cities' concentrator will start in 1974 and eventually will reach 2,000 tpd.

A considerably larger project is being developed by Cities Service at Pinto Valley, where a 40,000 tpd concentrator is being finished to process a 0.4% Cu and 0.01% Mo ore from a 350 million ton deposit. The plant will be in operation in 1974 and is scheduled to produce about 62,500 tons of copper per year. Concentrates produced in this plant will be smelted, as all other products from Cities Service plants, at Inspiration in Arizona.

Another new property ready to go by 1975 is the joint venture between Hecla and El Paso Natural Gas at Casa Grande. This $140 million project contemplates an underground mine which will produce about 15,500 tpd of ore, of which 9,000 tpd will be processed by flotation and 6,500 tpd by leaching and cementation. The concentrates from the flotation plant will be roasted, leached and electrolytically deposited. The roaster gases will be converted into sulfuric acid in a

250 tpd plant and used for leaching, while the calcines, high in iron, will be pelletized and used for copper cementation.

Phelps Dodge, in addition to its old operation at Morenci, which has a 120,000 tpy copper output, is now developing a new property at Metcalf, only a few miles to the north of the Morenci pit. This will be developed by 1975 into a 30,000 tpd milling operation with a 50,000 tpy copper output at a cost of about $180 million. This is a significant addition to the company's copper output, particularly taking into consideration that the expansion of the Tyrone operation in New Mexico was only finished in 1972, which raised the daily capacity of that plant from 30,000 tpd to 48,000 tpd and the copper output from 60,000 tpy to 100,000 tpy. With its properties at New Cornelia and Copper Queen still active and with the development of Safford, a large, 250,000 million ton 0.9% Cu orebody (not to be confused with Kennecott's low-grade property of the same name), this company is developing into one of the largest copper producers in the United States. Its present copper production capacity is already 330,000 tpy, and with new investments may soon reach 400,000 tpy, only one step behind Kennecott, so far the largest copper producer in the United States.

Inspiration, with its subsidiary at Christmas, continues to be the most integrated plant in Arizona. It uses three flowsheets for the treatment of its ores, including flotation and hydrometallurgical methods, recovers by-product molybdenite and processes its copper to refined and mercantile products such as rods. It also custom-smelts concentrates from other properties, for which reason it is actively engaged in pollution control studies.

A very interesting development has occurred at Bagdad which for many years operated a small but very efficient operation northwest of Phoenix. This company, which just recently introduced a unique process for high-purity copper production from cement copper (by using hydrogen precipitation) and which was one of the first to introduce ion exchange techniques for dump leaching practices, has now established that it is sitting on a 265 million ton orebody and wants to expand operations from a 6,000 tpd scale to 30,000 tpd. For this it needed about $100 million. The problem of financing was solved by merging with Cyprus, which already owns 50.01% of Pima and has some other mining interests.

In conclusion, it can be said that the southwestern American copper porphyries are fulfilling their goal of basic suppliers for American copper needs. Their capacity is rapidly expanding, and from the present level of 1,650,000 tons per year will reach 2,000,000 tpy by 1975 or 1976. Simultaneously a great deal of research and development work is being done for the solution of pollution problems, particularly through emission controls and the development of new processes. These problems, related to strict government regulations, at one time slowed down the development of new projects, but firm copper prices, greater demand for the red metal in this country and particularly foreign investment conditions and strategic considerations are gradually and confidently reversing this trend.

Chilean Production

In copper porphyries, Chilean copper output is next in importance to that of the United States as Table 4.3 shows. In recent years the Chilean copper industry has lived with a very difficult and complex situation caused by an extensive program for her expansion, political take over of this industry by Marxists during the short-lived Allende government, and an almost economic bankruptcy in which Chile found herself as a result of experimentation with "the road down to socialism."

While it is too early at the time of this writing (October 1973) to predict how fast the new approach of the military Junta will affect the recovery of the Chilean copper industry, it should be stressed that Chile has one of the most developed and solid raw materials bases, a relatively new and modern industrial structure, and uses one of the most advanced technologies in copper production in the world. All these factors are important prerequisites for a solid advance and future increase of Chilean copper output.

It should be stressed again that Chile is in possession of the world's largest copper reserves, some 90 million tons of metallic copper at this counting. This is almost one quarter of the world total, and incidentally these reserves have a great potential for growth. Secondly, Chilean copper ores are rich not only in copper but in molybdenum content as well. At the greatest Chilean copper deposits the ore mined always contains at least 1.7 percent copper and occasionally surpasses two percent. This is about 3 to 4 times more than in comparable deposits in the United States, Canada and Russia. Moreover, molybdenum by-product content in Chilean ore ranges anywhere from 0.03 to 0.05 percent Mo, which is at least two to three times above the average in other countries. Thus, the quality and quantity of Chilean copper ores are obviously a prime attraction to any investor.

Chileans have always realized that their copper potential was not fully used to the advantage of the country. As shown at the beginning

TABLE 4.3
PRODUCTION DATA ON LATIN AMERICAN AND CANADIAN COPPER PORPHYRIES

Name and Location	Ore production – tons		Heads		Annual Output – tons	
	Annual	Daily	% Cu	% Mo	Cu	Mo
24. Chuquicamata, Chile	20,000,000	60,000	2.00	0.037	300,000	2,500
25. El Teniente, Chile	20,000,000	62,500	1.85	0.04	280,000	2,400
26. El Salvador, Chile	8,000,000	24,000	1.35	0.024	100,000	1,500
27. Andina, Chile	3,500,000	10,000	1.60	0.015	65,000	–
28. Disputada, Chile	3,500,000	10,000	1.50	0.01	45,000	–
29. Mantos Blancos, Chile	2,200,000	6,000	1.80	–	36,000	–
30 Toquepala, Peru	15,300,000	40,000	1.10	0.018	150,000	750
31. Cananea, Mexico	8,000,000	24,000	0.70	–	45,000	–
TOTAL LATIN AMERICA	80,500,000	236,000	1.50	0.03	1,021,000	7,150
32. Lornex, B.C.	12,500,000	38,000	0.43	0.01	47,500	500
33. Gibraltar, B.C.	11,500,000	36,000	0.37	0.01	37,500	500
34. Bethlehem, B.C.	6,000,000	17,000	0.55	0.018	28,500	–
35. Island Copper, B.C.	11,000,000	33,000	0.52	0.017	50,000	1,250
36. Brenda, B.C.	9,000,000	24,000	0.18	0.03	15,000	3,750
37. Gaspe, Quebec	12,000,000	34,000	0.60	0.015	58,000	850
TOTAL CANADA	62,500,000	182,000	0.44	0.015	236,500	4,850

of this chapter, while the Russians exploit their copper porphyries at a 3.5 percent annual rate, and Canada and the United States at 1.5 to 2 percent, Chilean copper in the last few years was exploited at only 0.87 percent, and at much lower levels a few years ago.

While this fact can make some conservationists quite happy, it cannot be a positive sign for a nation which suffers from economic underdevelopment and which badly needs resources to develop her infrastructures. Some 90 million tons of copper and over 2 million tons of molybdenum in the Chilean subsoil represent today a market value of over $100 billion. This is too much for a nation which has only a $6 billion GNP and finds enormous difficulties in assembling a $2 billion annual budget. In other words, Chilean finances, economic development and general prosperity could be considerably boosted if, for instance, from a 0.87% annual exploitation rate Chile could reach a 2% annual exploitation rate, which is the world average. This fact alone could increase Chilean national income by $1.2 billion, or 20%, more than double Chilean exports and thus solve all national monetary problems and speed up industrial and economic development. For it should be clearly realized that mineral resources in the ground, unused and unexploited, are dead capital. To sit on them without using them is equivalent to sitting on a bag of gold and being hungry.

In these days of inflation and economic and political instability, many people invoke the idea that mineral resources in the ground are as good as money in the bank, or even better, because they do not undergo the same devaluation as money does due to inflation. Such opinions were recently heard among Arab oil producers. This theory may be true for rich people, who intend to transform their resources into money and keep it in the bank, but for nations who need cash for investment and development, such a position is equivalent to suicide. Everyone knows that wisely invested capital not only produces new jobs, reduces imports and offers more goods and services, but also doubles its value each seven to ten years. Such things cannot be expected from mineral resources in the ground, which in absolute terms cannot be expected to double in value each seven to ten years. In fact, it has often been observed that the prices of raw materials systematically grow slower than those of industrial products, which in effect means that raw materials producers each time can buy less for the same quantity of their raw materials.

In view of this it seems that maximum development of natural resources within the limits of national interest and ecology should be

Fig. 4.6 Chuquicamata Metallurgical Complex in Northern Chile.

a standing goal for developing nations in their progress toward industrial and economic development.

It seems that this type of thinking influenced the Christian Democrat President Eduardo Frei when in 1965, immediately after his election, he embarked on a scheme which was appropriately baptized "Chilenization" of Chilean copper and which essentially consisted of two trends: 1) a greater participation of the Chilean government and Chilean nationals in their copper business and 2) favorable capital and tax treatment for foreign capital which should provide funds for further drastic expansion of Chilean copper output. Originally it was thought to expand the existing Chilean copper output from an annual rate of 660,000 tons in 1965 to 1,100,000 tons by 1970 and probably as much as 1,200,000 tons immediately after, at an estimated investment cost of $700 million. The details of this plan are shown in Table 4.4.

The investment was made by 1970 and the first operations, such as those at Andina, El Teniente, Exotica and others, were put on stream by outgoing President Frei himself, when the elections of September 1970 installed as a new President the Marxist candidate Dr. Salvador Allende.

It would be unproductive at this stage to discuss the political background of copper in Chile. Being a principal industrial product of the country and accounting for roughly 15 to 20 percent of its Gross National Product and 85 percent of exports, this activity has always been considered to be of too crucial importance to be left to the management and direction of those outside the country. At one point President Frei himself expressed that: "A company owned abroad, and controlling resources upon which the entire life of a nation depends, is not compatible with the best interests of that nation."[15] Thus his "Chilenization" scheme always considered a gradual take over of foreign companies by the Chilean government in the long run. This process was speeded up by the incumbent Allende government, which made out of it a question of economic independence and national pride, and thus obtained the unanimous approval of all political parties in the Chilean Parliament for the nationalization of all large copper properties.

Circumstances, however, were considerably complicated when the American companies discovered that no compensation would be paid for the properties taken over and when the Marxist government started to politicize the mines. A massive influx of political partisans in the mines, replacement of able technical managers by political appointees, fragmentation of working discipline from above, strikes, and a number

TABLE 4.4

EXPANSION PROGRAM FOR THE CHILEAN COPPER INDUSTRY

Production Capacity

	In 1965	Planned for 1970	Cost 10^6
Chuquicamata	300,000	390,000	75.4
El Teniente	180,000	280,000	243.
El Salvador	100,000	110,000	12
Exotica	—	112,500*	48
Andina	—	65,000	157
Sugasca	—	12,000	33
ENAMI	40,000	97,000	85
Others	60,000	80,000	30
	680,000	1,146,500	683

* To be processed and refined at Chuquicamata oxide plant.

of technical problems related to the limited experience of new personnel, lack of spare parts and replacements, falling prices of copper, the economic and monetary difficulties of the country, eventually led to a disorganization of production processes, a fantastic increment in production costs and eventually to stagnation of output. In spite of built-in new additional installed capacity of some 300,000 to 400,000 tpy, the new managers never could essentially surpass levels achieved already in 1969 and 1970, when extra production capacity did not exist. This had a heavy impact on the national economy, because Chile was bearing the burden of a $700 million debt, acquired during the expansion of the copper industry, but did not obtain any extra copper to pay for it.

Following is the record of planned and achieved production during Allende's days:

	Planned*	Actual Production		
		1970	1971	1972
Chuquicamata	350,000	265,273	250,187	234,645
El Teniente	280,000	194,825	147,280	190,618
El Salvador	100,000	93,187	84,909	82,777
Andina	65,000	5,976	53,584	53,910
Exotica	112,500	4,034	35,264	31,271
Others	155,000	147,386	137,076	123,579
Total	1,062,000	710,681	708,300	716,800

* Installed capacity — all figures in metric tons.

The political abuse in large mines led to continuous strikes and the gradual deterioration of equipment and the lack of spare parts impacted production. Newly expanded El Teniente, which already in 1970 had achieved a 194,825 ton output, when new installations were not yet on stream, in 1971 experienced heavy losses due to strikes and production anarchy, and produced about ½ of its planned production. The expanded Chuquicamata mine never reached its planned 350,000 tpy production, but instead dropped its output to levels below the pre-expansion capacity. Having all the necessary smelter and refining capacity for the first time in history, it started to export concentrates and had to use smelters at Potrerillos and Paipote because of failures in its own smelter. The Exotica project never took off at the designed capacity because of gross miscalculation and the inadequacy of metallurgical processes. Only the Andina mine lived up to expectations, although on a somewhat lower level of production.

A typical example of production inefficiency was the rise in production costs due to inorganic incorporation of new workers and failing technology. For example, at El Teniente the production costs in November of 1970 were $0.21 per pound; by January of 1971 they rose to $0.34 per pound, and in May of the same year reached an unbelievable $0.52 per pound, 3 to 5 cents more than the market price.

At Exotica, in the 1971–1972 period, from the ore containing 150,000 tons of copper only 66,000 tons of copper were extracted and the production costs in 1972 reached an unbelievable 59.7 cents per pound or 20% above the market value of the product. On the whole, Chilean production costs in large copper mining, the most efficient sector of production even at artificial and manipulated rates of exchange, varied in the following way in cents per pound of copper produced:

Fig. 4.7 Exotica: a mine of many metallurgical problems.

1970	1971	1972
32.9	42.2	45.1

By 1973 the situation drastically worsened because of the galloping inflation, permanent strikes and production disorganization. Monthly production at largest mines fluctuated at Chuquicamata between 12,900 and 28,000 tons and at El Teniente from 7,800 to 20,500 tons, while smelters accumulated 200,000 tons of unprocessed or semi-processed materials. The overall efficiency in metal recovery declined from 80 percent in 1969, to 71 percent in 1971 and hit the bottom in 1973. The overall use of installed capacity fluctuated between only 73 and 75 percent.

The government in the beginning of 1971 predicted copper output for the same year at over 1,000,000 tons, only to finish the same year with a 708,300 ton production. Later, similar goals were never reached and the expected output for 1973 in August was estimated to have hit a low of 680,000 tons.

The situation started to improve rapidly after the military take over. The bringing back of discipline and capable management resulted in sharp increment of production almost immediately. Working under essentially the same conditions as in the last months of Allende, i.e., with shortages in supplies and spare parts, but now with cooperating and disciplined labor, production of big copper mining soared almost 50 percent. Andres Zaushquevich, the new executive vice president of CODELCO, the largest Chilean and world copper corporation, was able to announce last November that CODELCO output in October and November of 1973 soared to an average of 67,400 tons per month, as compared with an average of 45,000 tons per month during Allende's period.

Here is the comparative output of different mines of CODELCO during the military and Allende governments:

	October		*November*	
	1972	*1973*	*1972*	*1973*
Chuquicamata	15,544	28,349	21,195	29,707
El Teniente	14,924	20,120	15,816	21,321
El Salvador	7,207	7,970	7,003	7,206
Andina	4,143	6,032	3,094	6,026
Exotica	2,064	4,480	2,756	3,589
Totals	43,882	66,951	49,864	67,849

Fig. 4.8 New reverberatory furnace in construction at Caletones, El Teniente, Chile.

With this Chile has already embarked on a 900,000 tpy rate and is expected to reach 1,000,000 tons production in a year or so. There are some bottlenecks in smelting capacity at Chuquicamata, but they are expected to be solved by construction of a new smelter.

In general, Chile has a very promising future for expansion of copper production. Two of her giants, Chuquicamata and El Teniente, have ore reserves sufficient to keep an annual copper output of 500,000 and 750,000 tons per year each, as compared with only about 300,000 tpy rated capacity in each case at the present time; such production can be achieved with a relatively small additional investment because of the existence of all the necessary infrastructures. In other words, the extra capacity for 1 ton in annual copper production instead of the presently accepted $2,500 to $3,000 could cost only half or even less. At Chuquicamata, for instance, the concentrating capacity is about 85,000 tpd, while the actually run production is at 55,000–60,000 tpd. This is mainly due to the fact that at the present time ore heads run 2.5 to 3% copper and the plant is overcrowded with concentrates. Indeed a unique case, which would be so much appreciated in Arizona or British Columbia! With this in mind and with only minor adjustments in the mine and smelter, at a relatively low cost, Chuqui could rapidly build her production to 450,000 tpy and more. This would increase her present output by at least 50 to 60 percent.

Similar alternatives exist at El Teniente, where with a new reverberatory furnace in operation by early 1974, the smelter will be able to take all concentrates which two concentrators can produce. This will bring El Teniente to the designed 280,000 tpy capacity very rapidly. Future expansion plans may consider a new outlet to Codegua with a new industrial complex to be built only 50 miles from Santiago.

While new mining policies of the military government are not yet clearly spelled out, it is quite obvious that the Chilean government has decided to continue to run all CODELCO properties by itself and to invite participation of private capital in new ventures only. Such ventures will consider primarily development of new properties, such as are planned at El Abra, Los Pelambres, Anadacollo and others, although joint ventures in expansion of old properties may not be excluded. The government is working presently on setting ground rules for such investment which should benefit the national interest and be attractive to the potential investors.

Peruvian Production

In recent years, Peru has also shown a considerable interest in making better use of her copper resources and expanding her copper production. Until 1960 the average copper output of Peru was about 50,000 tons. It came from Cerro de Pasco and some smaller mines. However, in 1960 the large Toquepala copper porphyry was put into operation and production immediately zoomed to about 200,000 tons. In fact, in the last three years after the expansion of the Toquepala operation, it averaged some 235,000 tons per year.

The problem with Peru is the same as with many developing countries. It has natural resources but no development capital, and thus foreign capital is decisive for the increase of mineral output. Most of the Peruvian copper reserves are locked in large copper porphyries. The five largest of them contain some 2 billion tons of ore with 17 million tons of copper and close to 1 billion pounds of molybdenum. With such reserves Peru could easily produce 500,000 to 600,000 tons of copper per year instead of the present output of 235,000 tons. The main problem is in the attraction of investment capital.

Since its coming to power in 1968 the present military government embarked on a number of social reforms and introduced a new mining policy in accordance with some principles agreed upon by the Andean Pact. These policies call for a planned development of natural resources and have set a program for the fulfillment of exploration, feasibility and development studies. Most of the Peruvian copper porphyries were at that time in the hands of foreign companies which apparently were reluctant to make new investments in view of the uncertain political climate and the increased emphasis of the Peruvian government on the participation of workers in the ownership of the companies. To secure development of idle orebodies the new Peruvian mining law set schedules and dates for owners of mineral deposits to fulfill different stages of exploration and development studies and then to proceed with investment. Those who missed the schedule would lose their properties, which would then revert to the state. Under these circumstances, Anaconda lost its Cerro Verde and Santa Rosa properties, ASARCO its large Michiquillay deposit and Cerro de Pasco Tintaya, Antamina, Chalcobamba and Ferrobamba deposits. Southern Peru, which was the owner of Quellaveco and Cuajone, opted to lose Quellaveco and concentrate on Cuajone, a program which could find adequate financing only recently.

So far, the only copper porphyry in operation is the Toquepala mine, owned by Southern Peru, a concern consisting of four American companies. Its development made a landmark not only in Peruvian but in Latin American mining as well. Since the beginning of operations it already has produced close to 2,000,000 tons of copper and substantial quantities of by-product molybdenum. In recent years its profits were invested in the development of the Cuajone mine.

Now George Argall, Jr., a recognized authority on porphyry coppers, who previously extensively covered porphyry copper deposits around the world, including the U.S.A., Canada, Russia, the Fire Belt and Bulgaria in a recent issue of World Mining,[44] gives a detailed account of Chilean and Peruvian copper deposits and reports inherent difficulties — financial and technical — which should be met to develop a major property like Cuajone. He points out that only adequate copper prices may warrant huge expenditures for such new enterprises necessary under new circumstances.

The official government agency, Minero Peru, has recently published the following program for the development of the Peruvian copper industry:[16-17]

Project	Location	Annual Output	Project Cost	Expected Date
Cerro Verde	Arequipa	36,300	$ 88 million	1975
Cuajone	Ica	140,000	$355 million	1976
Tintaya	Cuzco	30,000	$ 48 million	1975
Santa Rosa	Arequipa	136,000: 1st stage	$ 70 million	1976
		168,000: 2nd stage		1977
Antamina	Ancash	60,000	$ 35 million	1977
Ferrobamba	Apurimac	150,000	$143 million	1976
Chalcobamba	Apurimac			
Michiquillay	Cajamarca	88,000	$400 million	1978
Quellaveco	Moquegua	66,000	$200 million	1977
Berenguela	Puno	33,000	$ 36 million	Not set

This $1.3 billion plus expansion program will obviously depend on the willingness of foreign banks to make loans or of investors to invest. Taking into consideration that these plans are parallel to similar ones for the development of the oil, metal-mechanical industries, power generation, and so on, which are also in billion dollar figures, it is hard to imagine that sufficient capital can be found in international markets for such dynamic developments. However, any progress toward

Fig. 4.9 Lornex Mining Corporation open pit at Highland Valley in British Columbia, Canada. (Photograph by George O. Argall, Jr. of World Mining.)

this 1,000,000 ton copper production goal will largely depend on the degree of confidence the Peruvian government can instill in investors. As the Peruvian Times recently commented,[18] "the international companies are interested in making profits, not in keeping already troubled subsidiary companies afloat" — which is largely the case for many multinationals today in Peru and some other Latin American countries.

Canadian Porphyries

With the exception of Noranda's Gaspe Mine in Quebec, all Canadian copper porphyries are in British Columbia. In recent years this region has been under intense prospection and exploration in view of very promising geological conditions for porphyry-type deposits. Both copper and molybdenum porphyries were discovered and several new mines put into operation. This has helped to almost double the 82,000 tpy output in the five years since 1968.

The oldest porphyry in British Columbia is Bethlehem Copper, which started its operation in 1962. It started to mine a property in Highland Valley, which according to some geologists may be one of the largest copper districts in the world, at a modest rate of 1.2 million tpy and gradually expanded this operation to 6 million tpy in 1972. Simultaneously it increased its copper output from 25 million pounds to 58 million in 1972. Today Bethlehem has one of the most solid mineral bases in British Columbia. It has leases, permits and reservations all over British Columbia and Alberta, and claims the following mineral properties:[19]

Property	*Ore reserves - tons*	*Average Grade % Cu*
Huestis and Jersey mines	38,500,000	0.56
J-A zone project	286,280,000	0.43, and 0.017% Mo
Lake zone project	190,000,000	0.48
Maggie ore zone	200,000,000	0.40 - equivalent
Total	714,780,000	0.44

With copper reserves of over 3 million tons of metallic copper, Bethlehem probably will soon expand her present operations which in 1972 amounted to only 29,000 tons. Bethlehem is 25% owned by Grangesberg of Sweden, and 24% by the Sumitomo Group of Japan. It's also considering, in conjunction with Noranda and Placer Development, the construction of the first copper smelter in British Columbia.

In 1972 three large mines in British Columbia were inaugurated, all with marginal copper content, but supplemented with substantial molybdenum by-product values. These three properties, Island Copper, Gibraltar and Lornex, are all based on large but low-grade reserves, and their success will greatly influence investment in such properties in the future.

Island Copper Mine of Utah Mines Limited, a wholly owned subsidiary of Utah International, an American firm, started its operations by the end of 1971. This company, located on the northern end of Vancouver Island, some 220 air miles from Vancouver, is based on a 280 million ton orebody averaging 0.52 percent copper and 0.025 percent molybdenum. The 33,000 tpd operation was constructed at an estimated cost of $70 million, and will produce 230,000 tpy of copper concentrates containing 50,000 tons of copper and some 1,800 tons of premium molybdenite concentrate, which also contains about 0.2 percent Re — the highest concentration so far found in the world. The value of rhenium can easily exceed that of molybdenum itself. This very modern operation smoothly entered into production without any major problems, and already is delivering its products to Japan, principally to Mitsui Mining and Smelting, which contracted for 60 percent of its output for the next 10 years, and to Mitsubishi Shoji Kaisha and the Dowa Mining Company, which will buy another 30 percent of production in the next 5 years.[20-21]

Another mine in operation since the second quarter of 1972 is Gilbraltar, a subsidiary of Placer Development, a company with international interests. Gilbraltar is located in the Cariboo District in central British Columbia, about 38 miles north of Williams Lake and 350 miles north of Vancouver. The operation is based on a 358 million ton orebody assaying from 0.37 to 0.45 percent copper and about 0.01 percent molybdenum. The 36,000 tpd operation was constructed at a cost of $67 million, and the annual copper output is expected to be 37,500 tons plus some 500 tons of molybdenum by-product.[22-23]

Lornex is the largest porphyry copper operation so far put on stream in British Columbia. It has a designed capacity of 38,000 tpd — some 12.5 million tons of ore per year — and an annual production of copper of 47,500 tons plus from 500 to 1,000 tpy of molybdenum as by-product. It was constructed at an estimated cost of $138 million, of which $91.6 million came from a consortium of Canadian banks and from Japanese clients.[24] To start its operation in the last quarter of 1972, Lornex had to remove some 47 million tons of glacial overburden and waste rock with some copper oxide values. Some initial difficulties were reported in operations, which in spite of a 10,450 ton copper and 581,000 pound molybdenum output and 87.9% copper recovery, left the company with a $1,215,000 net loss in the first three months of operation, up to the end of 1972.[25] Hopefully these difficulties are already overcome. Lornex is managed by Rio Algom Mines, a subsidiary of Rio Tinto-Zinc Corporation, and its future production is sold to Japan. It should be observed that Lornex is located in Highland Valley, next door to Bethlehem and in the middle of a cluster of copper porphyries in this area. The large Bethlehem J-A prospect is only 3 miles northeast of Lornex, the Cominco-Bethlehem Valley Copper prospect just 2.5 miles north of Lornex, and the prospective Highmond operation of Teck Corporation two miles southeast of Lornex.[26]

Brenda Mines, inaugurated only in 1970, is probably the lowest copper grade mining operation in the world. It is based on some 165 million tons of 0.2 percent copper and 0.05 percent molybdenum ore, and in the first three years of operation is expected to produce only 15,000 tons of copper and 7.5 million pounds of molybdenum per year. Brenda was constructed by Noranda Mines and loans from the Bank of Nova Scotia and Nippon Mining and Mitsui of Japan, at a cost of $60 million. Its success obviously will depend on the situation with the copper market, because the molybdenite market in recent years has been very difficult.[27]

Finally, with respect to Gaspe of Quebec, a Noranda subsidiary, it can be said that it was recently reconstructed and enlarged by a $85 million expansion program. Gaspe copper has reserves of some 230 million tons of ore assaying 0.4% copper and some additional molybdenum values. It was expanded three times to a daily concentrating capacity of 34,000 tons in the flotation part and to 5,000 tpd in leaching operations.[28] Eventually the copper output should reach 72,000 tpy, and sulfuric acid for leaching will come from a new acid plant at the enlarged smelter, which will use a continuous smelting process reactor.

Pacific Belt Porphyries

As indicated in Table 4.5, the Pacific Belt area processes today some 75 million tons of ore to obtain about 440,000 tpy of copper, ¾ million ounces of gold and 2.6 million ounces of silver. Almost half of this production comes from Bougainville Copper, a new production giant put into operation in April of 1972 at Panguna of Bougainville Island, Papua and New Guinea. This deposit was known since the 1930s, when it was worked as a gold mine. Further exploration has

TABLE 4.5
PRODUCTION DATA ON PORPHYRY COPPERS FROM THE PACIFIC AREA

	Ore production – tons		Heads			Annual Output – tons			
	Annual	Daily	% Cu	% Mo	Cu	Mo	Au oz.	Ag oz.
38. Bougainville, N. Guin.	32,000,000	90,000	0.74	0.004	200,000	—	500,000	1,300,000
39. Ertsberg, Indonesia	2,500,000	7,000	2.50	—	63,000	—	68,000	750,000
40. Biga, Philippines	11,500,000	31,500	0.38	—	30,900	—	20,000	120,000
41. Toledo – DAS, Philippines	12,800,000	34,500	0.47	—	50,100	—	22,500	137,000
42. Labo, Philippines	6,700,000	18,000	0.79	—	53,000*	—	51,000	213,000
43. Santo Thomas, Philip.	6,700,000	22,000	0.50	—	24,000	—	75,000	118,000
44. Sipalay, Philippines	3,500,000	14,500	0.74	0.01	20,000	100	—	—
TOTAL PACIFIC AREA	75,700,000	217,500	0.67	—	441,000	100	736,500	2,638,000

discovered large reserves of copper ore containing small amounts of molybdenum, gold and silver. At this writing, reserves are already reported to be [29] 900 million tons, assaying an average of 0.48% Cu, 0.004% Mo, 0.028 oz/t Au and 0.07 oz/t Ag. To produce some 200,000 tpy of copper the mine has to extract 165,000 tpd of ore and the mill process 90,000 tpd. The expected annual milling capacity is 32 million tons of ore and some 690,000 tons of copper concentrate assaying 28% Cu, 12% Mo, 26% Fe and 31% S, which will be sold principally to Japan and other copper consumers in the West. Bougainville is the property of Conzinc Rio-Tinto of Australia, which out of the total investment for its development of $436 million contributed $160 million, the balance being lent in Eurodollars from a group of banks and from Japanese copper clients, Mitsubishi Shoji Kaisha and Mitsui, who contributed $30 million in cash and another $30 million in Japanese equipment. The contracts in 1972 accounted for a minimum of 135,000 tpy for the next ten years.[30-31]

Another very interesting property is located at Ertsberg in the Indonesian part of New Guinea. This relatively small but high-grade orebody was known to exist for years, but because of difficult climatic conditions and political uncertainties stayed idle for many years. Now it has been developed at a cost of $150 million by Freeport Sulphur Company. It was inaugurated in March of 1972 and is planned to produce annually 63,000 tons of copper, 68,000 ounces of gold and some 750,000 ounces of silver. This property experienced some operational difficulties in the initial stages, but then the ore mined happened to be much richer than anticipated. Financing of this project was obtained from 30 sources and on the basis of contracts which call for delivery of $2/3$ of production to a group of 8 Japanese smelters and $1/3$ to Norddeutsche Affinery in Hamburg.[32]

The most active part of the Pacific Fire Belt, as far as copper production is concerned, is the Philippines. The copper output of this region increased in the last ten years some 3.4 times, from 70,000 tpy in 1963 to 236,000 tons in 1972. The share of five copper porphyries in this production was 196,000 tons, or 83 percent.

The most important copper producer in the Philippines is Atlas, which from its two porphyries at Biga and Toledo extracted last year some 80,000 tons of copper.

For this production it had to mill over 22 million tons of ore. Atlas has been immensely successful due to its rapid expansion of production. Its revenues in 1972 were close to $100 million[33] with a net income of $31 million. While in 1971 Atlas was 45% Filipino owned, in 1974

Fig. 4.10 Bougainville open pit started in 1971. (Photograph by George O. Argall, Jr. of World Mining.)

the national ownership will increase to 60 percent. Apart from gold and silver, Atlas also recovers as by-product a magnetite concentrate.

The second largest copper producer in the Philippines is Marcopper which operates a $200 million low-grade copper porphyry at Labo. In 1972 it processed 6.75 million tons and recovered about 50,000 tons of copper, 50,000 ounces of gold and 220,000 ounces of silver. Its revenues amounted to $48 million, of which $24 million was net income. The company has difficulties with the ownership since only 29 percent is owned by Filipino nationals.

Philex, the third largest copper producer, mined last year the same tonnage as Marcopper, but produced only half as much copper. The copper content of its Santo Thomas mine last year was only 0.375% Cu, but it recovered as by-product, gold, silver and magnetite concentrate. The total revenues in 1972 were $24.4 million of which net income accounted for a solid $13 million.[33]

Another very progressive Philippine company is Marinduque, owned by the Cabarrus family, which runs Sipalay Copper at Negros Occidental Province. The originally constructed 5,000 tpd mill was gradually expanded to 14,500 tpd and last year the company produced almost 28,000 tons of copper valued at over $22 million. In a recent report [34] the company claimed that its 2400 x 500 ft. oval orebody, which was drilled up to a depth of 2500 feet, contained 500 million tons of 0.5% Cu ore. At a depth of 2500 feet a 0.3% Cu material was still being found. Marinduque is also engaged in other metals business.

The Philippines, like British Columbia, has no copper smelter of its own. Recently a $100 million project was studied and presented to President Marcos by Lepanto Consolidated Mining Company. Lepanto is a medium-sized company with $46 million in revenue. It operates a small, high-grade property containing 2.7% Cu and produces some 25,000 to 30,000 tpy of copper along with 130,000 ounces of gold and 400,000 ounces of silver. However, its concentrate is high in arsenic and it had difficulties in shipping it to ASARCO for smelting.[35]

Communist Bloc Porphyries

Surprisingly, the copper production from porphyry coppers in the Communist World is still of relatively small importance. Out of an estimated copper output of 1,560,000 tons in 1972, only some 162,000 tons, or slightly over 10 percent came from porphyry coppers. This can be explained by two facts: first, copper porphyries found in the

TABLE 4.6

PRODUCTION DATA ON OPERATING COPPER PORPHYRIES IN THE COMMUNIST WORLD

Name and Location	Ore production – tons		Heads		Annual Output – tons	
	Annual	Daily	% Cu	% Mo	Cu	Mo
45. Balkhash, Kazakhstan	13,500,000	37,500	0.4	0.01	42,500	550
46. Bozshchakul, Kazakhstan	5,500,000	15,000	0.6	0.02	26,500	–
47. Almalyk, Uzbeckstan	18,000,000	50,000	0.7	0.01	100,000	720
48. Kadzharan & Armenia	7,000,000	20,000	1.2	0.05	60,000	1,650
49. Majdanpek, Yougoslavia	4,750,000	13,200	0.7	0.005	27,500	–
50. Medet, Bulgaria	8,000,000	26,500	0.35	0.008	27,500	320
Total Communist World	56,750,000	162,200			284,000	3,240

Communist World are generally very lean in copper content and very oxidized; second, there are still substantial reserves in vein copper, sedimentary deposits and others which can be conveniently processed.

So far only six large operations have been reported in production, four of them in the Soviet Union, and one in Yugoslavia and the last in Bulgaria, as indicated in Table 4.6.

The oldest and most important porphyry copper operation is at Balkhash. It was put into production in 1938, just before the war, and was the first large-size complex, later completely integrated, which served as a prototype for many other plants to be built. The Balkhash concentrator and smelter are in many ways similar to Kennecott's operations in Salt Lake City. They depend on the near-by open-pit mine at Kounrad, use Lake Balkhash salted water for mineral processing, and recover numerous by-products including molybdenite and rhenium. During World War II copper manufacturing was added to this integrated complex, which since that time has continued its development. The original medium-grade copper reserves from the Kounrad mine are already mined out, and the remaining low-grade material, apparently cannot supply enough feed for a 37,500 tpd concentrator. For this reason a new mine was developed at Saiak, at the eastern end of Lake Balkhash, and put into operation a couple of years ago. Balkhash has smelting and refining facilities well in excess of 100,000 tpy, probably as much as 150,000 tpy. However, its own concentration plant supplies only a modest 40,000 to 50,000 tpy of copper equivalent. Molybdenite recovery comes from two sources:[36] as by-product recovery from copper concentrates and from a straight treatment of a molybdenite ore found near the Kounrad orebody. This molybdenite, as well as some other concentrates, are processed for rhenium recovery.

The Bozshchakul mine and mill are an old project which took over 10 years to develop. It is a low-grade and very oxidized orebody with insignificant molybdenite mineralization. The principal problem for treatment of this ore was lack of water. This was solved by the construction of the Karaganda canal which brought water from the Siberian river, Irtysh. Concentrates obtained at Bozshchakul are shipped for smelting and refining to Balkhash.

Almalyk is a relatively new operation, inaugurated in 1958 as a part of the Altyn-Topkan combine and later expanded to its present 50,000 tpd size. It impresses everyone by its largeness, and no accurate estimate of its capacity is available. However, it is known that this mill has been expanded several times and the flow-sheet has been

TABLE 4.7
PROPHYRY COPPER PROJECTS IN DEVELOPMENT

Property & Location	Property of	Planned Invest. in $ million	Milling TPD	Annual Cu output tons	Expected Date of iniciation
1. Pinto Valley, Arizona	Cities Service		40,000	62,500	1974
2. Lakeshore, Arizona	Hecla	140	Sulphide 9,000 oxide 6,500	50,000	1974 1974
3. Miami, Arizona	Cities Service	100	expand to 40,000		1974
4. Black Sea Copper, Turkey		100	40,000		1973
5. Bethlehem	Bethlehem Copper		25,000		1975
6. Bagdad Copper	Cyprus Mines	100	expand to 30,000		1975
7. Mamut, Malasia	Japanese Dev.	80	15,000	30,000	1975
8. Twin Buttes, Arizona	Anaconda-Amax	200	expand to 40,000 & leach 10,000	100,000 30,000	1975
9. Cerro Verde, Peru	MINEROPERU	88		30,000	1976
10. Cuajone, Peru	Southern Peru	355	40,000	130,000	1976
11. Cerro Colorado, Panama	Canadian Javelin	560	175,000	400,000	1980
12. Pachon, Argentina	Aguilar-St. Joe	200			
13. Krivelj, Yugoslavia	Yugoslavia	200	60,000	60,000	
14. Metcalf, Arizona	Phelps Dodge	180	30,000	50,000	1975

changed as well. It tried the LPF process, which eventually failed, and apart from copper recovers a relatively small amount of by-product molybdenum.

Kadzharan with its 20,000 tpd capacity is the largest Armenian plant. This plant, built immediately after the Second World War, treats a relatively rich ore containing 1.2% Cu and 0.05% Mo from a nearby mine of the same name. Other, smaller plants at Agarak, Alaguez and Dastakert process similar ores, but on a smaller scale. The most modern of them is at Agarak, which was built in 1963. It is located on the Iranian border. A copper concentrator exists also at Kafan. All concentrates are smelted in the large, modern smelter at Alaverdy.

The Majdanpek mine, near the large copper producer at Bor, has its own concentrating plant built in 1962 according to western technological standards and with western equipment. Its ore is very lean in molybdenum, so only copper is recovered in this 13,200 tpd operation at 27,500 tpy. Now, very near to Majdanpek, a new but very low-grade porphyry has been discovered at Krivelj.[37-39]

Bulgaria's Medet was built with Russian help and with Russian technology. This mine also produces 27,500 tpy of copper in spite of the fact that it mines and mills double tonnage in comparison with Majdanpek. It has been described in detail by several authors, a relatively unusual fact for mines in the Communist World.[40,41,42]

Projects under Development

In Table 4.7, some of the major projects presently in the development stage are given. It can be observed that the 4 to 5 percent annual increase in copper demand requires each year an approximate addition of about 400,000 tons in new copper at an investment cost of over $1 billion. With the increasing insecurity of foreign investment all around the world, a greater emphasis by industrial nations is being placed on domestic copper, or investment is shifted to very remote areas where the danger of nationalization is less. To obtain an approximate 1,000,000 tons of copper in the next two to three years, at least a $2.5 billion investment will be necessary.

BIBLIOGRAPHY FOR CHAPTER FOUR

1. R. U. King, D. R. Shave and E. M. MacKevett, Jr.: Molybdenum; United States Mineral Resources, Prof. Paper 820, 1973, pp. 425-436.

2. Andrew Kuklis: Molybdenum; USBM Yearbook for 1970, pp. 727-736.
3. Kennecott Copper Corporation Annual Report for 1972.
4. Alexander Sutulov: Molybdenum and Rhenium Recovery from Porphyry Coppers; University of Concepcion, Chile, 1970.
5. USBM Yearbook for 1970, p. 1219.
6. USBM Yearbook for 1970, p. 468.
7. John V. Beall: Copper in the US — a Position Survey; Mining Engineering, April 1973, pp. 35-47.
8. Newmont Annual Report for 1969.
9. Esperanza and Ithaca Peak Ore Grades, Reserves and Costs; World Mining, March 1969, p. 82.
10. The Significance of Sierrita; Metal Bulletin, London, October 23, 1970, pp. 24-28.
11. Duval Sierrita Mills 71,252 tpd; World Mining, June 1972.
12. Anaconda's Annual Report for 1969.
13. An ex-banker treats copper's sickest giant; Business Week, February 19, 1972.
14. Statement of Consolidated Income for 1971 by Anaconda Company, February 10, 1972.
15. Latin America — No Longer a US Sphere of Influence; US News & World Report, December 20, 1971, pp. 77-80.
16. The Slow Scramble for New Project Finance; Peruvian Times, April 14, 1972; Finance Marketing Survey, p. 20.
17. Good Prospects for Future Mining Exports; Peruvian Times, August 10, 1973, p. 3.
18. Finance and Marketing Survey; Peruvian Times, April 14, 1972, p. 21.
19. Bethlehem Copper Corporation 1971 and 1972 Annual Reports.
20. Island Copper on Schedule; Mining Journal, August 20, 1971, p. 153.
21. Island Copper Project; Mining Magazine, October 1972, pp. 344-349.
22. Gibraltar Copper Mines Starts; World Mining, July 1972, pp. 42-45.
23. Canada's Newest Copper Mine Now in Production; American Metal Market, June 21, 1972, p. 13.
24. Lornex Strips 173,000 tons, builds 38,000 tpd mill; World Mining, November 1971, pp. 58-59.
25. World Mining, March 1973, p. 102.

26. Lornex; Mining Magazine, March 1973, pp. 154-163.
27. Brenda Stripping on Schedule; World Mining, July 1969, p. 23.
28. Noranda Still Relies on Mineral Industry; American Metal Market, March 16, 1971.
29. R. W. Ballmer: Personal communication.
30. R. W. Ballmer: The Bougainville Copper Project; Mining Congress Journal, April 1973, pp. 33-40.
31. George Argall: Bougainville Ships Ahead of Plan; World Mining, October 1972, pp. 45-76.
32. Ertsberg Copper Mine Officially Opened; Mining Magazine, April 1973, p. 221.
33. Philippine Mining Record, June 1972.
34. Enrique Bugarin: Personal communication.
35. Philippine Mining Record, February 1973.
36. Alexander Sutulov: Molybdenum and Rhenium Recovery from Porphyry Coppers; University of Concepcion, Chile, 1970.
37. Alexander Sutulov: Mineral Resources and the Economy of the USSR; McGraw-Hill, New York, 1973.
38. A. W. Knoerr: Majdanpek; Engineering & Mining Journal, September 1962, pp. 77-88.
39. Yugoslavia's Krivelj Project; Mining Magazine, May 1973, pp. 324-355.
40. George Argall: Bulgaria's Medet; World Mining, October 1970, pp. 32-37.
41. Dimiter Derlipanski: Bulgarian Non-Ferrous Metals; World Mining, February 1973.
42. P. T. Shirkov et al.: Industrial Development of Low Grade Porphyry; VIII International Mineral Processing Congress, Moskva, 1968.
43. Harold J. Bennett: An Economic Appraisal of the Supply of Copper from Primary Domestic Sources. US Bureau of Mines Information Circular No. 8598, 1973.
44. George O. Argall, Jr.: Cuajone — developing the world's next copper porphyry mine; World Mining, November 1973.

Fig. 0.5 A modern concentrator of porphyry copper ores at Butte, Montana.

CHAPTER FIVE

Metallurgy of Porphyry Coppers

Most copper porphyries are worked today by open pit mining, because this permits the use of larger and more efficient equipment and eliminates some of the hazards and costs of protection.* In such leading copper producers as the United States and the Soviet Union, open pit mining predominates over underground methods at least in the proportion of 3:1, and in 1970 some 84 percent of copper produced in the United States was from open pit mines, while in the Soviet Union all porphyry copper with the exception of the Armenian mines was also obtained from open pit mines. Only when the ore deposit is too deep, or when the climatic conditions do not favor open pit mining, are underground methods applied, generally block caving. Such is the case with San Manuel, El Salvador, El Teniente and several others.

Crushing and Grinding

The first step in any ore beneficiation scheme is size reduction, in order to achieve liberation of mineral values from gangue minerals for further processing. It is used both in flowsheets which contemplate concentration by flotation and in hydrometallurgical flowsheets when straightforward leaching is contemplated.

Due to the inflationary pressures and higher cost of capital which increasingly press upon costs of production, in the last decade or so there has been considerable emphasis placed on increasing the size of equipment and simplification of flowsheets in order to maintain production costs within reasonable limits. Another factor, of course, was the

---

* A recent U.S. Bureau of Mines study indicates that typically a 30,000 tpd. open pit operation will require a capital investment of about $10 million while the underground block caving mine of the same production capacity may require $63 million. The respective operating costs are $0.40 and $2.45 per ton of ore.

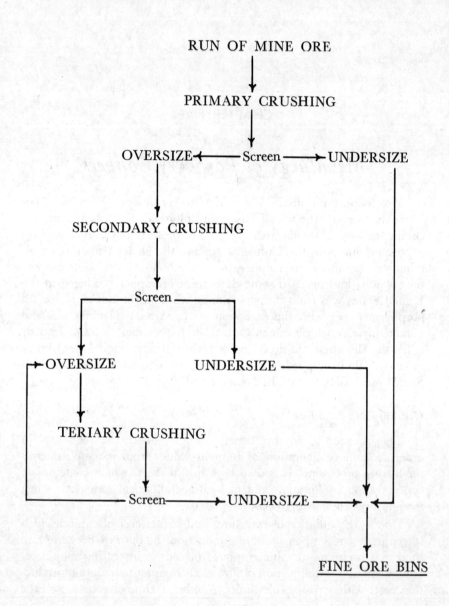

Fig. 5.1 Typical Flowsheet for a Porphyry Copper Crushing Plant

ever decreasing grade of our ores and their poorer metal content. If Daniel Jackling, in the beginning of this century, should have proved those ores containing as much as 40 pounds of copper per ton to be commercial ores, today we would have reached the stage when ten pounds per ton was a standard case and at times would have to satisfy ourselves with a 6 to 7 pound copper equivalent.**

As is known, size reduction is one of the costliest beneficiation operations because of the expensive equipment involved, high consumption of energy and media for grinding as well as the wear-out of equipment. The cost of crushing and grinding in overall ore treatment normally accounts for 40 percent of all ore dressing expenses, but occasionally may go as high as 50 or more percent. No wonder, then, that crushing and grinding along with their auxiliary operations have been the subject of careful studies in the last few years.

A typical flowsheet for a copper porphyry crushing plant can be appreciated from Fig. 5.1. Generally, it can be stated that the present tendency is to locate the crushing stage as close to the mine as possible and thus simplify all the problems of materials handling and transportation. As a general rule it is accepted that the crusher should be able to take any rock which the mine delivers, i.e., which the shovel can manage; and gyratory crushers with their large openings for the feed have definitely won predominance over other types of crushers in the primary crushing stage. As Table 5.1 shows, all of the newer crushing plants, and for that matter even the older ones, invariably have a large gyratory crusher for their primary crushing step.

The secondary or intermediate stage of crushing commonly employs seven foot Standard Cone Crushers, while the tertiary, fine crushing stage generally uses Short Heads, also of cone type. It is obvious that the modern tendencies in contemporary milling are best observed from the design of the latest plants, such as Twin Buttes, Pima and Sierrita in the United States; Lornex, Gibraltar, Island Copper and Brenda in Canada; Andina and the new El Teniente plant in Chile; Bougainville in Papua and New Guinea. They are as indicative for the seventies as Mission, Esperanza, Mineral Park and the new mill at Butte were for the sixties.

If previously two or three stage crushing followed by two stage grinding was standard, today this scheme is rapidly changing. In some cases a three stage crushing is followed by a single stage ball milling, as is the case at Sierrita and Bougainville, or at El Teniente, where there

---

** I.e., combined value of copper and by-products expressed in pounds of copper.

TABLE 5.1

SIZE – REDUCTION EQUIPMENT USED IN MODERN PORPHYRY COPPER CONCENTRATORS

Name of Plant	Date of inauguration	CRUSHING				GRINDING		
		Primary	Secondary	Tertiary	Prod.	Primary	Secondary	Overflow
1. Sierrita	1970	58"x89" Gy	7' HC	7' HC		16½'x19' B		
2. Twin Buttes	1970	30"x70" Gy	7' Std	7' SH	5/16"	14'x18½' R	14½'x28' P	20" Cy 53%-200m
3. Pima	1972*	54"x74" Gy				28'x12' Aut.	16½'x19' B	20" Cy 62%-200m
4. Lornex	1972	60"x89" Gy				32'x15½' Aut	16½'x23' B	20" Cy 55%-200m
5. Gibraltar	1973	54"x74" Gy	13"x84" HC		6-8"	13½'x20' Rd	13½'x20' B	24" Cy 75%-100m
6. Bethlehem	1962	42"x65" Gy	7' Std	7' SH		12½'x15' Rd	11'x14' B	20" Cy 65%-200m
7. Island Copper	1972	54"x72" Gy			9"	32'x14' Aut		20" Cy 65%-200m
8. Bougainville	1972	54"x74" Gy	7' Std	7' SH	½"	18'x21' B		20" Cy 42%-200m
9. El Teniente	1970**	underground	7' Std	7' SH		14'x24' B		24" Cy 67%-200m
10. Andina	1970	underground	7' Std	7' SH		11'x16' R	12'x16' B	15" Cy 75%-200m

* After fourth expansion to 53,000 tpd
** New 27,500 tpd mill constructed for expansion of operations

Abbreviations:
Gy - gyratory crusher
HC - hydrocone crusher
Std - Standard cone crusher
SH - short head cone crusher
RD - rod mill
B - ball mill
Aut - autogenous mill
Cy - cyclone

is only two stage crushing because of underground mining which by itself eliminates the necessity for primary crushing.

In other schemes, a revolutionary advance has been made by autogenous grinding, which like Pima, Lornex or Island Copper reduces crushing to only one, primary stage, with 8 to 9 inch discharge being immediately fed to large, 32 foot diameter mills, driven by 4,000 horsepower motors. Such a combination results either in the production of a finished product for flotation, as found at Island Copper, or requires some additional pebble or ball grinding, as used at Lornex and Pima.

A 54 to 60 inch Primary Gyrator is quite standard equipment today for primary crushing. Seven foot Symons Standards and Short Heads are standard for secondary and tertiary crushing, and a ¾ to ⅝ inch product is an average crushing plant discharge, except for the cases when autogenous grinding is applied. This equipment is normally operated by intermediate 500 horsepower to 1,000 horsepower motors.

In an attempt to reduce construction costs and to make operations more flexible, particularly in regions with a good climate such as Arizona and Uzbeckstan, large, intermediate storage piles have replaced the traditional ore bins. They not only eliminate much of the heavy construction costs for silos, but permit a round-the-clock crushing plant operation, if desired, while the mine can stay on a five-day operation basis. Also, repairs and maintenance work can be conveniently made without loss of production. Such piles can contain 150,000 and 200,000 tons of ore and can be conveniently evacuated by conveyors installed in concrete tunnels under the stockpile and fed by slot feeders.

Another modern tendency both in crushing and grinding plants is the replacement of bridge cranes by heavy-duty mobile cranes, which makes considerable savings in construction, particularly by eliminating the necessity for expensive supporting columns and roofs.[41]

In fine crushing, horizontal arrangements with all crushers on a single floor have definitely won predominance,[42] and the tertiary crushing step is often closed with a second set of screens in order to insure granulometric characteristics of fine-ore bins product.

In grinding, the predominant philosophy is to direct hard ores for autogenous grinding while reserving a single stage ball milling for softer ores.[41] Since its successful application at Esperanza, single stage ball milling has become increasingly popular where lower capital cost is sought, as, in the cases of Sierrita and Bougainville. In the case of El Teniente, it was applied in spite of the fact that a similar ore was being two stage ground in the older mill.

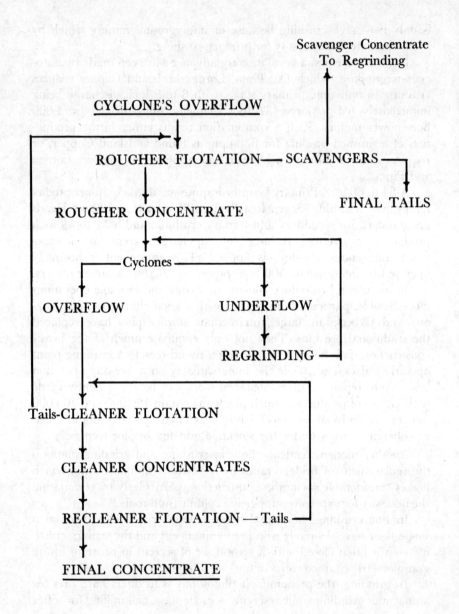

Fig. 5.2 Copper Flotation Circuits in Porphyry Copper Plants

If two stage grinding is applied, a rod mill-ball mill combination is still very common; two ball mills, of essentially the same size as the rod mill, follow each rod mill because of the approximately double quantity of work input to be done in secondary grinding as compared with the primary stage.

Primary rod mills generally work in an open circuit, their discharge being scalped for fines by cyclones, and the secondary ball mills, as is obvious, always work in closed circuit, cyclones being generally the apparatus producing the final overflow. Cyclones are increasingly replacing traditional classifiers, particularly in fine classification, with great advantages in equipment cost, space and operational costs. Introduced after the Second World War, they produced quite a revolutionary change in the design of grinding floors in the sixties, drastically reducing space and investment.

Grinding mills today use motors anywhere from 1,500 horsepower for conventional grinding rod and ball mills up to 4,500 horsepower for autogenous grinding.

Flotation Technology

The secondary cyclone's overflow, which is the flotation feed, normally goes to the flotation plant first, for rougher and then for cleaner flotation, as indicated in the flowsheet shown in Fig. 5.2. This flowsheet is rather standard today and is used almost in all copper producers. If the final concentrate contains recoverable molybdenum values, before final thickening and filtering the cleaner concentrate goes to a retreatment plant.

The cyclone's overflow is generally already conditioned with the necessary flotation reagents and lime, for a pH which can vary anywhere from 9 to 12, depending on the presence of pyrite and the necessity of its depression. A factor is also the presence of by-product molybdenum in the ore, because molybdenite recovery into flotation concentrate is the best at pH between 9 and 10.

If all of the flotation reagents are not added to the pulp in the grinding circuit, they are added before the rougher flotation step either to a conditioner or in the first cells of the flotation machines.

The only substantial change in the flotation process in recent years except for different flotation formulae and automatic controls is the considerable increase in the size of flotation equipment. This has grown from 50 and 100 cubic feet in capacity to 200 and 300 cubic feet and is already reaching 500 cubic feet. In the not so distant future it may

Fig. 5.3 Large flotation cells produced by Denver Equipment Co.

grow even further. If the brand new 30,000 tpd mill at San Manuel in 1956 needed 480 flotation cells for rougher flotation (four sections, with six banks of machines, each containing 20 cells of 48 inch Agitairs, with a 357 tph output per section), the 1973 mill at Gibraltar, with a 20% greater capacity, employs in the same circuit only forty-eight 300 cubic foot 600H Denver flotation cells (three sections, with one bank of 16 flotation cells).

Other modern plants, such as Lornex, with a 38,000 ton daily capacity inaugurated in 1972, use four banks of eight 300 cubic foot Denver 600H for rougher flotation, and another four section of 10-cell identical machines for scavenging. Island Copper, with 33,000 tons daily capacity and 21 minutes of rougher flotation and scavenging time, uses for this purpose 10 banks of 14 cell 300 cubic foot machines.

Such a drastic increase in the size of flotation cells along with a substantial decrease in the number of processing sections in the mill are plainly justified by substantial savings in equipment costs, construction and foundation costs, as well as in piping, electrical and control equipment and instrumentation. Also operational costs can be reduced somewhat. Shoemaker and Taylor,[41] for instance, claim that in a 20,000 tpd flotation plant by only using the largest available equipment as compared with conventional plants, some 16 percent of costs could be saved on account of equipment and construction savings, while another 16 percent could be saved if some other modern features of design were incorporated. Here is the breakdown of their costs:[41]

TABLE 5.2

COMPARATIVE COSTS FOR A 20,000 TPD
COPPER CONCENTRATOR
(expressed in millions of dollars)

	Conventional	Large Equipment	1970 Design
Yard & Nonprocess	4.8	4.8	4.8
Coarse Ore	4.0	4.0	3.7
Fine Ore	5.1	4.6	3.8
Concentrator	15.5	11.1	7.2
Process Water	0.3	0.3	0.3
Reclaim Water	0.7	0.7	0.7
Tailings Disposal	0.6	0.6	0.6
Total Project	$31.0	$26.1	$21.0
$ per daily ton treated	1,550	1,300	1,050
Percentage of conventional	100	84	68

TABLE 5.3
METALLURGY OF OPERATING COPPER PORPHYRIES

PLANT	HEADS		ROUGHER CON.		% RECOVERY		CLEANER CON.		MOLY CON.		% Mo Rec.	% OVERALL RECOVERY	
	% Cu	% Mo	% Cu	% Mo	Cu	Mo	% Cu	% Mo	% MoS$_2$	% Cu		% Cu	% Mo
1. Magna & Arthur	0.69	0.03	27	0.9	90	80	30	0.12	90	1.0	70	90	56
2. San Manuel	0.69	0.015	28	0.5	92	80	30	0.06	92	1.0	86	92	70
3. Butte	0.76	—	14	—	—	—	26	—	—	—	—	85	—
4. Morenci	0.83	0.015	14	0.1	—	—	22	0.14	Production curtailed			75	—
5. Sierrita	0.29	0.03	7	0.7	86	75	26	0.15	87	2.5	90	85	67
6. Pima	0.56	0.017	26	0.2	86	64	28	0.15	64	2.1	71	85	45
7. Tyron	0.89	0.013	12	—	—	—	22	0.15	—	—	—	78	—
8. Twin Buttes	0.6	0.03	12	0.6	86	50	29	0.15	73	1.1	70	76	35
9. Ray	0.98	0.015	18	0.2	82	50	18	0.05	86	1.0	75	82	37
10. New Cornelia	0.70	—	14	—	—	—	30	—	—	—	—	85	—
11. Yarington	0.5	—	15	—	—	—	30	—	—	—	—	84	—
12. Chino	0.90	0.008	20	0.23	79	70	22	0.06	85	1.0	57	79	40
13. McGill	0.85	0.016	19	0.2	78	30	20	0.05	60	1.5	50	78	15
14. Inspiration	0.71	0.007	35	1.0	76	50	38	0.17	95	0.4	60	76	30
15. Mission	0.7	0.02	25	1.0	90	75	28	0.1	87	0.3	90	89	67
16. Mineral Park	0.42	0.03	18	1.3	76	65	20	0.04	90	0.3	95	76	62
17. Copper Queen	0.83	—	4	—	—	—	10	—	—	—	—	75	—
18. Esperanza	0.37	0.03	22	1.3	87	84	25	0.12	96	0.2	88	87	74
19. Miami	0.50	0.005	27	0.2	87	60	27	0.08	89	1.0	90	87	54
20. Silver Bell	0.7	0.008	9	0.15	86	50	30	0.18	85	0.8	60	84	30
21. Bagdad	0.7	0.03	32	1.5	88	80	32	0.15	90	1.0	80	88	64
22. Christmas	0.80	—	—	—	—	—	20	—	—	—	—	72	—

23. Battle Mountain	0.84	—	—	—	—	—	—	—	—	—
24. Chuquicamata	2.00	0.037	50	0.70	91	50	25	0.2	92	—	69	—
25. El Teniente	1.85	0.04	20	0.3	83	60	50	0.15	95	0.3	90	40
26. El Salvador	1.35	0.024	46	0.8	81	75	40	0.1	97	0.5	83	42
27. Andina	**1.90**	0.015	28	0.2	**89**	—	48	—	—	0.2	81	65
28. Disputada	1.50	0.01	—	—	—	—	28	—	—	—	**90**	—
							25				85	—
						Leaching Operation					
29. Mantos Blancos	1.68	—	—	—	—	—					90	—
30. Toquepala	1.20	0.018	31	0.3	88	62	32	0.12	87	1.2	88	37
31. Cananea	0.70	—	—	—	—	—	30	—	—	—	**88**	—
32. Lornex	0.43	0.01	33	0.3	88	80	34	0.1	84	1.0	88	64
33. Gibraltar	0.37	0.01	30	0.7	88	60	30	0.15	85	1.0	85	45
34. Bethlehem	0.55	0.018	12	—	—	—	32	—	—	—	87	—
35. Island Copper	0.52	0.017	5	0.13	89	75	24	0.67	90	0.7	88	67
36. Brenda	0.18	0.03	22	3.0	90	85	25	0.2	91	0.1	88	81
37. Gaspe	0.60	0.015	27	0.3	90	70	30	0.06	90	0.5	90	56
38. Bougainville	0.74	0.004	13	—	—	—	24	—	—	—	85	—
39. Ertsberg	2.50	—	—	—	—	—	28	—	—	—	93	—
40. Biga	0.38	—	9	—	—	—	29	—	—	—	79	—
41. Toledo	0.47	—	—	—	—	—	28	—	—	—	81	—
42. Labo	0.79	—	—	—	—	—	25	—	—	—	85	—
43. Santo Thomas	0.38	—	—	—	—	—	24	—	—	—	90	—
44. Sipalay	0.74	**0.015**	26	0.7	86	50	27	0.1	90	0.8	85	30
45. Balkhash	0.4	0.01	12	0.2	85	55	18	0.05	60	NA	85	41
46. Bozshchakul	0.6	0.02	—	—	—	—	17	—	—	—	75	—
47. Almalyk	0.7	0.01	15	0.2	83	50	17	0.04	50	NA	83	40
48. Kadzharan	1.2	0.05	10	1.0	72	60	16	0.2	82	0.7	70	47
49. Majdanpek	0.7	0.005	—	—	—	—	25	—	—	—	84	—
50. Medet	0.35	0.008	14	0.4	77	60	25	0.1	70	NA	76	52

109

In other words, up to ⅓ of the investment cost could be saved by using very large equipment, reducing the overall height of construction, eliminating bridge cranes and replacing them with heavy-duty mobile cranes, replacing silos with stockpiles, etc.

Rougher Flotation

In Table 5.3 a summary of metallurgical data for all 50 operating copper porphyries is given. Some 28 of them recover molybdenite as a by-product, and data on cleaner and molybdenite circuits are given as well.

Generally, it will be observed that the mill heads are normally subjected to a rougher flotation in which a rougher and less pure concentrate is collected and a general mill tail discharged. The grade of a rougher concentrate will obviously depend on its mineralogical composition, on the concentration ratio and on recovery. Thus from heads which normally vary between 0.5 and 1.0 percent copper and 0.01 to 0.03 percent molybdenum, concentrates assaying from 10 to 30 percent copper and from 0.2 to 1 percent Mo are obtained. These are then reground, refloated in two or more cleaner circuits and cleaner concentrates assaying up to 25 to 50 percent copper and 0.5 to 2 percent molybdenum are obtained with recoveries which for copper vary between 80 and 90 percent while for molybdenite they are somewhat lower, between 60 and 80 percent. If the recovery of by-product molybdenite is economically warranted, then it is carried out by processes which will be discussed further.

The general objectives of rougher flotation can be stated as follows: recover as much copper as possible into primary concentrate so that the tails discharge contains as little as possible copper values. This step is thus complemented by a scavenging operation, the main objective of which is to reduce copper losses. Stimultaneously with copper recovery a maximum molybdenite recovery is required, but not at the expense of copper metallurgy. In other words, molybdenum recovery is secondary in importance to copper metallurgy. Only when the initial ore content in the molybdenum is high may optimization studies take place. Finally, in this first step of flotation, elimination of as much pyrite as possible is desired; thus, a preference for an alkaline circuit over the acid one. The only case of the acid circuit used in a copper porphyry is that of El Teniente in Chile, which is justified by additional copper recovery.* Pyrite, which floats readily in the acid circuit, is later eliminated in a retreatment process in the alkaline circuit.

The economic importance of primary copper recovery is obvious. At a 10 million tpy rate and 1 percent copper heads, each percent in copper recovery means 1,000 tpy of copper, i.e., about $1,000,000 — thus the constant research for better reagents and flowsheets which will enhance better copper recoveries.

Collectors

The spectrum of flotation reagents used today in primary copper flotation is very wide and ranges from traditional xanthates and aerofloats to more sophisticated synthetic products such as dithiocarbamates and methyl-isobutyl-carbinol.

In Table 5.4 a general survey shows the flotation reagents used in rougher circuits for some 40 operating copper porphyries. It will immediately be noticed that all of these properties, regardless of copper mineralization, in all instances use an alkaline circuit for primary copper and molybdenum recovery. Even at El Teniente, the acid circuit is used only for ores mined from upper levels of the mine, while the ore mined from the lower, primary zone shows better recoveries in the alkaline circuit. The pH of the alkaline circuit varies anywhere from an almost natural 8.5 to the highly alkaline 12, when difficult problems of pyrite depression are encountered.

A reasonable liberation for primary recovery is normally achieved at about 50 to 60 percent minus 200 mesh, although exceptionally coarse grinds of 33 and 40 percent minus 200 are not unusual. The concentrates from this float will be normally reground to 90 to 100 percent minus 200 mesh, at which the optimum liberation for copper and molybdenum values is reached and no obstacles for their effective separation are expected.

The preferred collector for primary flotation is considered to be xanthates, because of their flotation efficiency and low cost. In 80 percent of the cases they are used as the principal collector, i.e., in 32 of 40 flotation plants. Sometimes two different xanthates complement the optimum collecting action. Dixantogens, known under the name of Minerecs, are used as principal collectors only in 4 cases, or 10 percent of the total. They are absolutely indispensable in the acid circuit because xanthates will decompose in that media. On the other hand,

---

* This has not been found true, though, for lower level ores, which are processed in a new plant in an alkaline circuit.

TABLE 5.4
PRIMARY FLOTATION OF PORPHYRY COPPER ORES

Plant	Predominant Mineralization			Grind 2-200m	pH	Collectors						Frothers				
	Chalcopyrite	Chalcocite	Cpy/cte			Dithiocarbonate	Dithiophosphate	Dixantogene	Dithiocarbamate	Hydrocarbons	Others	Pine Oil	Cresylic acid	MIBC	Dowfroth 250	Others
Utah	X	–	–	60	8.5	–	X	–	–	S	–	–	X	X	–	–
San Manuel	–	–	X	60	10.5	X	–	X	–	C	X	–	–	X	–	–
Butte	–	X	–	50	10.5	X	–	X	X	–	–	–	–	X	–	–
Morenci	–	X	–	52	10.5	–	–	X	–	–	–	–	X	X	–	–
Sierrita	X	–	–	55	11.0	X	–	–	–	–	X	–	–	X	–	–
Pima	X	–	–	62	11.5	X	–	–	–	X	–	–	–	X	–	–
Tyrone	–	X	–	60	11.4	X	–	–	–	–	–	–	–	X	–	–
Twin Buttes	X	–	–	53	11.0	X	–	C	–	–	–	–	–	–	–	X
Ray	–	–	X	60	11.5	X	–	–	–	X	–	–	–	X	–	–
New Cornelia	X	–	–	43	10.5	X	X	–	–	–	–	–	–	–	–	X
Yerington	X	–	–	50	11.0	X	–	–	–	–	–	X	–	–	–	X
Chino	–	–	X	55	11.0	X	–	–	–	X	–	–	X	–	X	–
McGill	X	–	–	52	10.8	X	–	–	X	X	–	–	X	–	–	–
Inspiration	–	X	–	55	10.5	X	–	X	–	–	–	–	–	X	–	X
Mission	X	–	–	55	11.5	X	–	–	–	–	–	–	X	X	–	–
Mineral Park	–	X	–	60	11.5	X	–	–	–	X	–	–	–	X	–	–
Copper Queen	–	–	X	50	11.5	X	–	X	–	–	–	–	–	–	X	–
Esperanza	–	–	X	60	11.5	X	–	–	X	X	–	–	–	–	–	X
Miami	X	–	–	35	11.0	–	X	X	X	–	–	X	–	–	X	–
Silver Bell	–	–	X	65	11.0	–	X	–	–	–	–	–	X	X	–	–
Bagdad	X	–	–	40	11.5	X	–	–	–	X	–	–	X	–	–	–
Christmas	X	–	–	50	9.5	X	–	–	–	–	–	–	X	–	–	–
Chuquicamata	–	X	–	46	11.0	X	X	–	–	–	–	–	X	–	X	–
El Teniente	–	–	X	67	4.2	–	X	–	–	–	–	–	–	–	–	X
El Salvador	–	X	–	66	11.0	X	–	X	–	–	–	–	X	–	–	–
Andina	X	–	–	70	9.0	X	–	–	X	–	–	–	–	X	–	–
Toquepala	–	X	–	60	11.5	X	–	C	–	–	–	–	X	C	–	X
Lornex	X	–	–	55	9.5	X	–	–	–	–	–	–	X	–	X	–
Gibraltar	X	–	–	65	10.5	X	–	X	X	–	–	–	–	X	–	–
Bethlehem	X	–	–	64	10.5	X	–	–	–	X	–	–	–	X	–	–
Island Copper	X	–	–	68	10.5	X	–	–	–	–	–	–	–	–	–	–
Brenda	X	–	–	40	8.0	X	–	–	–	X	–	–	–	X	–	–
Gaspe	–	–	X	55	10.0	X	–	–	–	X	–	–	–	–	–	X
Bougainville	X	–	–	50	8.5	X	–	–	–	–	–	–	–	X	–	–
Biga	X	–	–	50	8.5	X	X	–	–	–	–	–	–	X	–	–
Sipalay	X	–	–	33	7.8	X	X	–	–	–	–	–	X	–	–	–
Balkhash	–	–	X	62	12.2	X	–	–	–	S	–	–	–	–	–	X
Almalyk	X	–	–	46	11.5	X	–	–	–	S	–	–	–	–	–	X
Majdanpek	X	–	–	50	10.0	–	X	–	–	–	–	–	–	–	X	–
Medet	–	–	X	50	10.0	X	–	–	–	–	–	X	–	–	–	–

C - Cleaner circuit   S - Scavenger

the dithiophosphates, known as aerofloats and priced for their selectivity, are not so popular as they used to be and account today for only 9 out of 40 of the described operations. In some cases they serve as a complementing collector. Dithiocarbamate, known under the name of Dow Z-200, has so far made an entry into only 7 operations and is used as the only collector in 2 properties, as a complementing collector in 4 properties and in the cleaner circuit only in one property. There are also a number of other collectors in use such as S-3302, which is an alyl ester of amyl xanthate, and hydrocarbons, which promote molybdenite flotation.

Xanthates have proven to be excellent collectors for all types of principal mineralization found in copper porphyries starting from primary chalcopyrite and continuing through the secondary chalcocite, including mixed chalcopyrite-chalcocite ores. The most popular xanthate appears to be butyl xanthate, which is used in almost 50 percent of ores treated. It is particularly popular in the Soviet Union and Bulgaria where it is used almost exclusively as a sole collector. The other widely used xanthates are iso-propyl and amyl xanthates. The uses of ethyl xanthate are very limited and generally confined to use with other collectors.

As indicated before, flotation of copper minerals is often stimulated by complementary collectors, which are aerofloats, S-3302 and Z-200. The S-3302 collector fortifies not only copper flotation but that of molybdenite as well. The use of hydrocarbons for additional molybdenite recovery occurs in 12 out of the 28 known operations. It ranges from fuel oil to straightforward kerosene, and should be handled carefully in order not to affect the froth. A long conditioning in grinding circuits is thus favorable. When froth conditions are nevertheless affected, hydrocarbon collectors are added either to scavengers or to cleaners. In the latter case, of course, they do not affect rougher flotation but are instrumental to better metallurgy in the secondary circuits.

Wada and his collaborators have studied the effect of kerosene on primary molybdenite flotation.[43] They have found that conditioning prior to flotation improves selectivity with respect to pyrite. Also, distilled fractions of kerosene in the intermediate boiling range have the best properties, which also improve by mixing kerosene with pine oil. The use of emulsifying reagents also improves molybdenite recovery. The non-ionic emulsifying reagents, particularly polyglycolesters of fatty acids or alkylophenol polyglycolethers, gave good results, improving considerably the deteriorated froth conditions. In general, in molybdenite flotation with the help of hydrocarbons the exact proportion be-

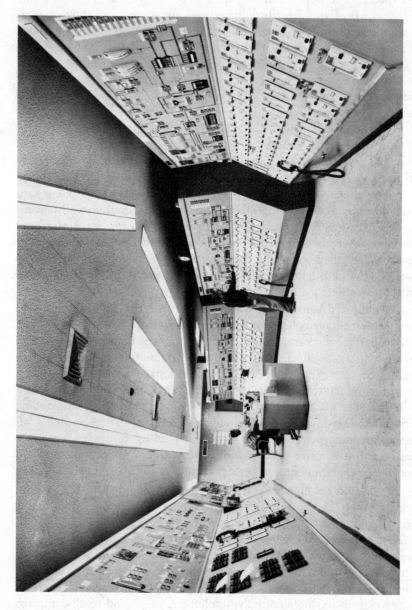

Fig. 5.4 Control room of a modern flotation concentrator.

tween hydrocarbons and frothers should be studied for each case in order to establish optimum froth conditions and to avoid froth degradation and metallurgical losses.

The alyl ester of amyl xanthate — S-3302 — is generally used in neutral or low alkaline pH values. It permits the elimination of much lime since, in contrast to xanthates, it is not a strong collector for pyrite and is helpful in the subsequent copper-molybdenum retreatment and separation.

Frothers

With respect to the frothers, attention should be paid to the growing use of Methyl-Isobutyl-Carbinol (MIBC), an alcoholic frother which received great popularity particularly in the United States. Just a few years ago pine oil, a natural and cheap product, undisputably dominated the flotation scene in copper metallurgy. Today its uses are confined to $\frac{1}{3}$ of all porphyry copper operations, generally in older mills, while MIBC already commands close to $\frac{1}{2}$ of all porphyries and is increasingly popular in new operations. It was used right from the beginning in many modern mills such as Sierrita, Pima, Tyrone, Gibraltar, Brenda, Bougainville, etc., and also made inroads into older operations such as Utah Copper, Butte, Ray and McGill after much testing and consideration. In some cases MIBC action is conveniently complemented by other frothers, like pine oil or cresylic acid. This is the case at Arthur and Magna, McGill, Mission, Silver Bell and Bagdad.

The use of this and other alcoholic frothers is particularly popular when hydrocarbons are added for stronger promotion of molybdenite flotation, because apparently hydrocarbons do not influence frothing properties of MIBC and other alcohols very strongly. This is not the case with natural products such as cresylic acid and pine oil. Some years ago, a new alcoholic frother known under the name of Powell was successfully introduced at El Teniente, where it replaced cresylic acid. It is not uncommon, however, to combine a synthetic and natural frother for optimum frothing conditions, the first generally being a measurable reference factor with respect to the second. This is particularly the case at some new Canadian and older American mills. The Russians still stick with natural frothers probably because of their cheapness and ready availability. Natural frothers should be used, however, in higher concentration than the synthetic ones. While their average addition ranges from 0.1 to 0.4 lbs/t, in the case of MIBC the average in the United States is only 0.06 lbs/t.

Fig. 5.5 Two internal photos of Medet flotation plant in Bulgaria, by George O. Argall, Jr.

Russia has developed several frothers of her own.[44] In rougher copper-molybdenum flotation heavy pyridine, a by-product of coke plants, is very popular. It is an alkaline tertiary amine, which due to contamination with components and impurities has a dark, oily aspect. It is not water soluble, but mixes readily with other oils. It is used in 75 percent of copper-molybdenum operations at about 0.35 lbs/t.

Another popular Russian frother is OPSB, which is a mixture of monobutyl esters of polypropylenglycols and which is prepared synthetically from propylen oxide and butyl alcohol. This is used mainly in Armenian plants.

Conditioning Reagents

Rougher flotation is normally carried out in an alkaline circuit. This is basically necessary for pyrite depression and for better copper and molybdenite recovery. Also, in an alkaline circuit much of the metallic ions are eliminated from the pulp by precipitation, which cleans the water and reduces reagent consumption.

Rougher flotation is normally carried out with an alkalinity which varies from 8.5 to 12, depending on the pyrite problem. If the pyrite is fresh and well crystallized without superficial oxidation, pH should be high. If the concentration of pyrite is low, or it is affected by oxidation, pH could be as low as 8.5. However, in the greatest number of cases pH would normally fluctuate between 10.5 and 11.5.

To produce the desired alkalinity, an average of 4 lbs/t of lime is consumed. It has been claimed that high alkalinity may affect molybdenite flotability and that the optimum molybdenite flotation takes place at pH 8.5. However, cases are known where molybdenite floats well at pH 10, 11 and even 11.5. All depends, apparently, on the type of mineralization and particularly on the presence of fine slimes. Lime is known as a powerful flocculator, and in the case of the presence of sericitic and kaolinic slimes, may affect molybdenite recovery by flocculating these slimes and causing the depression of finely disseminated molybdenite particles. This, then, is the reason why in some cases sodium hydroxide or soda is recommended as an alkalizer. Sodium hydroxide, apart from being an alkalizer, is an efficient dispersant. However, no large operations have as yet substituted lime for other alkalizers, apparently because of high costs, at least in rougher flotation.

In two cases rougher flotation has been carried out in the acid circuit. The specific reason for this is the rather oxidized nature of the treated ores. One case is at El Teniente, where it was proved that the acid circuit gives better copper recoveries than the alkaline, because it helps to clean some tarnished particles and recover some extra oxide and sufide copper. The molybdenite recoveries in the acid circuit are affected only slightly, probably no more than 2 or 3% in comparison with the alkaline circuit.

The other case was at Ray, Arizona, where the LPF Process was used to treat a part of the oxidized copper ore. Here flotation is also carried out at pH 4 and no particular influence on molybdenite recoveries has been observed.

It is estimated that today about 95% of the tonnage treated in rougher copper-molybdenum circuits uses lime as a conditioning reagent.

Another important conditioning reagent, principally used in Russia, is sodium sulfide. As is known, Russian copper ores are considerably more oxidized and affected by secondary changes than the ores of Arizona or Latin America. This implies, then, the use of sodium sulfide as a sulfidizer of copper ores. The quantities are, of course, low — about 0.4 lbs/t — in order to avoid the depression of copper sulfides. In the West, however, an acid circuit and LPF techniques have been used; for instance, at El Teniente, Morenci, Bagdad and Inspiration. Only at Chino is sodium sulfhydride used as a sulfidizer. The oxidation problem of Russian ores is also responsible for the higher consumption of collectors, which is about two times higher than in the West.

When pyrite depression cannot be effected efficiently with an alkaline circuit, small quantities of sodium cyanide are added to fulfil this purpose. The added quantities should be very small, on the order of 0.02 to 0.05 lbs/t, in order not to depress copper and molybdenite flotation.

Another Russian finding is that copper sulfate is a molybdenite promoter. At Balkhash it is added in the cleaner flotation circuit to promote molybdenite recovery. Apparently similar practices exist at Mission in Arizona. Kurochkina and Mitrofanov, using radioactive tracers, have proved that cupric ions absorb on molybdenite and increase the fixation of xanthates and aerofloats.

Crowding Out Effect and Lag

As we have already mentioned, molybdenite, in spite of its high flotability, quite often develops a lag in flotation in copper circuits. As Herlund[45] points out, in some cases molybdenite floats worse than quartz, and the problem of molybdenite recoveries should be considered in light of the fact that about half a pound of moly should be obtained from 2,000 lbs of ore, parallel to the recovery of a dozen or so pounds of copper, without upsetting the metallurgy of the latter.

The molybdenite lag in flotation can be understood in view of two facts: first, although the normally used ionic collectors float moly, they perform this flotation with less efficiency than the flotation of other copper and iron sulfides, as explained before; and second, because of the crowding out effect. Naturally, a third possible factor could be sliming, oxidation and other problems which may affect copper and molybdenum minerals equally.

The crowding out effect was interpreted by Beloglazov[46] many years ago. According to this author, the selectivity index of two minerals is determined by the following formula:

$$s, = \left(\frac{\partial_1}{\partial_2}\right)^2 \left(\frac{n_1}{n_2}\right)$$

where ∂_1 and ∂_2 are flotation activities
and n_1 and n_2 the number of particles in the pulp.

If we take into consideration that the number of copper and pyrite particles is from 200 to 500 times higher than that of molybdenite, it is obvious that even the highest difference in flotability between molybdenite and other minerals cannot compensate for the tremendous difference in the number of particles.

In industrial practice it is often observed, however, that the problematic and lagging particles, even if recovered, are dropped out in cleaner or recleaner circuits. This indicates that lagging is not only due to crowding out, but also to other phenomena.

Relationship between Mineralization and Reagents Use

Although this problem is extremely difficult to analyze with any clarity in respect to the actual mine mineralization because of its complexity, we will do it in view of the existing statistical material.

In the first place, we should stress that the mineralization of different ore-bodies as indicated in Table 5.4 is defined as predominant. This means that although some mineralogical components may represent a principal part of the ore, nevertheless other mineral components may be present in important quantities. Thus two minerals classified in the same group may vary considerably, not only with respect to the gangue components, but even with respect to the presence of other copper minerals.

Secondly, we should state that in many cases the ore is not mined from the same place, level, or even from the same mine. Quite often it is blended or not, according to necessity, and this certainly changes its metallurgical characteristics.

Finally, as the mining goes on, important parts of the orebody and types of the ores are eliminated. In a classical case of an open pit mine we may pass through the mined sulfide-oxide ore, reach the secondary zone and then spiral down to the primary zone. All this means strong variation in copper mineralization and should be taken into account. Thus the observations will be necessarily very general and will refer mostly to average ore types of known ores.

The most uniform condition can be observed in Russian mines. All of them are predominant in chalcopyrite mineralization and only one, Balkhash, has a mined chalcocite-chalcopyrite mineralization. Also, in all Russian mills a unique collector-butyl xanthate is used. Also, most plants help the collecting action of xanthate with hydrocarbons, but these are added in different points of the circuit depending on local conditions. It seems that where the flotation plants vary is in pH of the circuit and in type of frother used. It seems that Russian plants use mostly natural and readily available products from chemical or other industries. Thus, typification does not necessarily mean the optimum metallurgy, but may be just the readily available solution of a problem.

Things are rather different in the West where over 100 different flotation reagents produced by several big chemical concerns make a very competitive market. The western metallurgist, apart from the greater variety of products offered, has also greater liberty in introducing new reagents in plant practice, if economically justified. Thus, quite frequently changes of reagents with consequent improved mill metallurgy and the appearance of new and more competitive reagents are the results.

In the case of chalcocite-chalcopyrite predominating mineralization, which is apparently the most common, it seems clear that in all cases xanthates are used as principal collectors. Frothers may vary, but pine oil and MIBC strongly hold the market. The predominant pH of the circuits will be between 11 and 12 in all cases with lime.

With respect to the primary chalcopyrite or chalcopyrite-bornite flotation, xanthates are also predominant collectors with the exception of Utah Copper where dicresyl-dithiophosphate is traditionally used.

Only in predominantly chalcocitic mineralization is the use of xanthates not so predominant. Out of six properties, only four use xanthates, but there are two with aerofloats and one with Minerec and the other with Z-200 exclusively.

Cleaner Flotation

The rougher concentrates are a crude product which contain the bulk of recoverable copper and molybdenum values. They are still sufficiently coarse to contain a lot of middlings both with gangue minerals and between copper and molybdenite. Thus they should be, first, ground to a finer mesh size, and then upgraded.

Theoretically, copper-molybdenum separation from porphyry ores could be done in four ways:

1. Differential flotation of molybdenite prior to copper flotation into a rougher concentrate.
2. Differential flotation of molybdenite after rougher copper flotation.
3. Separation of molybdenite from the rougher copper concentrate.
4. Separation of molybdenite from the cleaner copper concentrate.

The first method was never used, because it would require a lot of reagents to keep cover values depressed while floating molybdenite. Also, molybdenite values in the ore should be very high to justify such a differential step. The second method was originally used at Cananea in Mexico, where for the first time molybdenite by-product was recovered from a porphyry copper in 1933. However, it was a special case which was justified by the high molybdenum content of the heads which was up to 0.5 percent. It has not been used since that time. Separation of molybdenite from a rougher concentrate can be effected efficiently only after its regrinding, or after sand-slime separation, as in the flowsheet used at El Teniente. However, this method is not very practical or popular. This leaves the fourth method as the only one which is almost exclusively used for copper-molybdenum separation. First, upgrading of the copper concentrate is done with a carefully controlled liberation of the copper and molybdenum values and the processing of this concentrate by one of several patented methods which will be discussed later.

For the time being it is important to observe that as a general rule, rougher copper concentrates undergo a regrinding step as shown in Fig. 5.2, and then are refloated in a cleaner and a recleaner circuit with or without additional reagents.

It is important to note that regrinding and liberation operations should be carried out very carefully because of the natural softness of molybdenite and its tendency for sliming. The sliming of molybdenite is closely related to the "edge effect," the significance of which will be understood from the explanation given in the next chapter, and which basically consists of the following: molybdenite consists of hydrophobic flakes which are hydrophobic on their face surfaces where the sulfur atoms are exposed. However, the edge parts of flakes are normally hydrophilic due to the fact that they represent broken bonds still active for reaction with water. If molybdenite is ground very finely, the edge surface increases in relation to the hydrophobic plate surface, and this makes the molybdenite particles more hydrophilic or less flotable.

This, then, is the reason why regrinding of rougher concentrate is kept strictly to the necessary minimum, in order not to reduce the flotability of molybdenite. Thus, as shown in Fig. 5.2, rougher concentrates are first passed through cyclones to eliminate fines and thus avoid overgrinding, and only coarse sands are reground. These are reground in a closed circuit with the same cyclones, so that continuous elimination of the liberated product is assured as fast as it is being produced.

The overflow from the cyclones then goes into a cleaning and recleaning circuit which consists of machines smaller than rougher flotation cells. For instance, if today 300 cubic foot to 500 cubic foot flotation is used in a rougher flotation, in a cleaner circuit 100 to 120 cubic foot flotation cells will be used. Recleaner circuits will normally consist of the same or even smaller sized cells. Tails from the cleaner and recleaner circuits are recirculated to the heads of the previous circuit.

The final copper concentrate, depending on the character of the mineralization and its grade of purity, will normally contain between 25 and 50 percent copper and 0.5 and 2 percent molybdenum, and is a product ready for molybdenite by-product recovery if this is commercially advantageous. Otherwise, if moly is not recovered, it will go for thickening, filtering, drying and smelting.

Special Processes

One of the most serious problems in the primary recovery of porphyry copper minerals is partial oxidation. The oxide copper minerals would not normally float with the same efficiency as copper sulfides. If sulfide copper minerals normally float with an efficiency of over 90 percent, the non-sulfide or oxide copper minerals, as they are called, recover generally into copper concentrate with an efficiency lower than 50 percent. Thus partially or heavily oxidized copper ores may experience substantial recovery losses in a rougher flotation circuit.

To give a few examples: a relatively clean sulfide copper ore unaffected by oxidation as in the cases of Magna and Arthur mills in Utah, or at San Manuel in Arizona, or at Brenda in British Columbia, will over yield a plus 90 percent copper recovery and sometimes 92 and 93 percent recovery, even if the heads are very low as in the case of Brenda. On the other hand, average copper ore from Arizona, such as at Pima, Twin Buttes, Esperanza, Miami, Bagdad, or El Teniente in Chile, to name a few, will always have a partial oxidation problem. The total oxide copper content of these ores may vary from 5 to 10 percent of the total copper values and this fact will be generally responsible (the other factor is insufficient grinding) for lower than normal copper recoveries. In this case if 5 percent copper oxidation lowers general copper recoveries by 3 or 4 percent, to an 86 to 88 percent level, a 10 percent oxidation may be responsible for recoveries as low as 83 and 85 percent.

When oxidation is higher than 10 percent, copper recoveries start to deteriorate quite considerably. The clearest examples of this are at Ray, Chino, and McGill, where partial oxidation of ores is responsible for recoveries which fall even below the 80 percent recovery level. It can be predicted almost unmistakably that if recoveries of copper in the primary flotation are below 80 percent, a difficult oxidation problem in some form or another must be present. Such is the case at Kadzharan and various Russian copper porphyries, at Medet in Bulgaria, at El Salvador and Chuquicamata in Chile and at many other places.

There were several lines of approach available to solve this problem, and they differed substantially depending on the degree of oxidation.

In the case of a slight, superficial oxidation, that of El Teniente, pH of the circuit was changed for improved copper recoveries. By introducing an acid circuit, part of the oxide copper, which covers sulfides as films, is dissolved, achieving two advantages: 1) the oxide coated sulfides are superficially cleaned and activated for a better flotation, and 2) the dissolved oxide copper values precipitate on the grinding media and are recovered into rougher concentrate as a highly flotable cement copper. This solution, which at El Teniente improved primary copper recoveries from 2 to 5 percent as compared with the alkaline circuit, did not work in the case of Chuquicamata or El

Salvador, where alkaline circuits always offered better recoveries, probably because of difficult slime problems.

LPF Process

If the oxidation of copper ore is such that at least 2 or 3 pounds of copper are lost per ton of ore because of poor flotation of oxides, stronger metallurgical action is warranted. We refer to the LPF Process.

In its original form the LPF Process was first suggested by Harmer Keyes in 1929 after his research on beneficiation of highly oxidized sulfide ores at Miami Copper in Arizona. A few months later, in 1930, the Russians claimed an independent discovery of the same process while studying the metallurgical problems of Kounrad ores for their future plant at Balkhash. In Russia this process is known under the name of the Mostovich Process, in remembrance of Professor V. Y. Mostovich and I. N. Duhanin, who tested Kazakhstan ores in Tomsk, Siberia. The third original investigator who should be considered is J. L. Stivens, who in 1932 patented the LPF Process in its contemporary form for Kennecott under U.S. patent 1,848,396.[47]

Essentially, the LPF Process is based on the following fundamental facts:

1. In a mixed sulfide-oxide copper ore, non-sulfide copper values tend to report in finer fractions than sulfides, because of their lesser mechanical resistance to crushing and grinding processes and strong sliming properties.
2. This permits, by a simple sand-slime separation operation, the obtaining of two products, where sulfides with very small fractions of oxides will report essentially in the sand fraction, while most oxides and finely dispersed sulfides will appear in the slime fraction.
3. The sand fraction, normally recovered from the primary classifiers, after grinding to about 10 mesh in primary mills should be reground in a secondary grinding circuit to the necessary granulometry, and then floated in a conventional sulfide circuit.
4. The slime fraction is first leached with sulfuric acid, by which oxide copper values go into solution, and the remaining sulfides are cleaned by acid in their surfaces; then, finely divided sponge iron is added to the pulp by which dissolved copper is precipitated as cement copper; and then the sulfides and cement copper are jointly floated either in an acid or alkaline circuit in a conventional way. These three consecutive steps — Leaching, Precipitation and Flotation — give the process its name by their initials.

It can be observed that the fundamental fact which underlies the LPF Process is that the leached copper values can be cemented in the pulp and floated together with sulfides into a copper concentrate. In this sense, sand-slime separation is an auxiliary operation to improve the efficiency and to reduce acid consumption. It is obvious that the same treatment which is applied to slimes alone could be used for the entire ore — and this is done in some cases when acid cleaning of sulfides in sands and some additional leaching of the copper oxides found in this fraction seem advantageous.

The acid leach is generally carried at pH 1.5 to 2 with a quantity of acid which may vary from a few to a dozen pounds per ton. The necessary quantity of acid will obviously depend on the quantity of copper to be leached and on the acid consumption by gangue and other minerals. This is not an acid leach in the full sense, but merely an acid wash, which will take care of relatively easily leached copper non-sulfides in a 10 to 15 minute lapse of time with intensive agitation, which is generally done in the flotation cells themselves. The stequiometric considerations require 1.5 units of acid per 1 unit of copper, but in practice this consumption is from 2 to 3 units of sulfuric acid per unit of leached copper.

Cementation of copper is normally done by sponge iron, which is a finely dispersed powder of metallic iron produced either from iron ores or from pyrite. The advantage of this product is that it has a large surface and rapidly precipitates copper by simple agitation in the pulp, which can be also carried either in flotation cells or special agitators. However, since the production of satisfactory sponge iron sometimes encounters difficulties at copper plants, finely ground shreds or iron pellets have been used in some cases, particularly in Russia, where this process has never been very successful. The theoretical consumption of metallic iron per unit of cement copper produced should be only 0.88 units of iron per unit of copper, but in practice is also considerably larger, from 1.5 to 2 units because of secondary reactions. Precipitation should be carried out at pH 3 to 3.5 and iron consumption can be reduced by shortening the precipitation time to a minimum, which is about 10 minutes. The excess sponge iron in the pulp can be recovered by magnetic separation and thus recirculated. This type of iron should not be exposed to air because of the fast oxidation rate.

Finally, the flotation step is carried either in the acid or alkaline circuit, or both when sand-slime separation is applied. In the alkaline circuit the best flotation results are obtained with xanthates and pine oil. In an acid circuit Minerec is normally used as the best collector.

In this case pH of flotation is adjusted to about 4 in order not to affect frothing conditions.

An integrated LPF plant should be able to be autonomous in acid and sponge iron supply. This is achieved by special attention to recovering a pyrite concentrate which will serve both for sulfuric acid production and sponge iron fabrication. The process is as follows: pyrite, as a harder mineral, is normally segregated into coarser, sand fraction. It is depressed in sulfide copper flotation with lime, but then reactivated with acid from tails and recovered into a clean concentrate by successive flotation steps up to the point when, say, an 80 to 85 percent pure product is obtained. This product is then fed into a fluo-solid reactor, which produces a high concentration SO_2 gas used for sulfuric acid production and discharges a calcine, which is conveyed into a reduction rotatory furnace when sponge iron is produced under controlled conditions. The final sponge iron product, which should contain at least 50 percent metallic iron, can be upgraded by magnetic concentration, which eliminates gangue minerals and unreacted iron oxides.

In spite of its elegance and simplicity the practical application of the LPF Process found considerable difficulties, particularly in the preparation of cheap and effective sponge iron product and the protection of equipment from corrosion. Then contamination of pulps with different metallic ions, due to the presence of sulfuric acid in the pulp, also affected in some cases the flotation properties of some copper minerals. All this on several occasions made the process expensive and uneconomical and was a direct reason for its suspension. Today in Russia LPF has been altogether abandoned after many unsuccessful runs at Almalyk and other places. It does not operate, either, at Ray, where a modern plant was constructed in 1958 and where a classic flowsheet was used.[3-5] It operates, though, at Butte, although on a limited scale.[48]

When used at Ray, the LPF Process showed the following advantage over conventional sulfide flotation:[49]

	% Cu Heads	% Cu Conc.	% Cu Tails	% Cu Recovery
Conventional	1.16	28.36	0.42	65.1
LPF Process	1.15	27.23	0.19	84.4

The non-sulfide copper content of heads was 0.34 percent, which largely reported in tails in a conventional scheme. By introducing the LPF Process in this case 4.6 extra pounds of copper were recovered per ton of ore, which plainly justified the new investment required, protection against corrosion and the additional consumption of acid and sponge iron.

The LPF Process was widely investigated and reported in the technical literature. Just recently an excellent survey on this process appeared in the Mineral Science and Engineering Review Journal.[50] The latest application of the LPF Process was made at Morenci.

Copper Hydrometallurgy

When the oxide copper problem is difficult and cannot be solved by conventional flotation or modified flotation processes like LPF, hydrometallurgical treatment of copper ores is necessary. This can range from straightforward leaching in vats, as is used at Yerington or Chuquicamata, to the application of some combination of techniques like the Dual Process at Inspiration. The increased attention to copper recovery from old dumps and tailings disposal has introduced a number of new and interesting alternatives which will be discussed in this chapter as well.

A classical case for a leaching operation is Yerington.[28] This plant, constructed about two decades ago, consists of a crushing plant which reduces all ore to minus 3/8 inch, which is then processed either by leaching or by grinding and flotation. Leaching is carried in 8 tanks each of 12,300 tons capacity. The tanks are constructed of reinforced concrete and lined with asphalt mastic. The ore is bedded in the tanks in six layers, each about 3 feet thick, in order to avoid segregation of material and channeling of solutions or blocking of circulation. Efficiency of leaching depends on the acid solution strength, which at the beginning is 30 to 35 gpl and by the end of the cycle about 1 to 2 gpl. The leaching time in each tank is about 120 hours, which results in a 92 percent oxide copper recovery and 86 percent overall copper recovery. Ore assaying 0.81% Cu contains about 0.68 acid soluble copper, and the tails discharged contain about 0.12% Cu, half in sulfide and the other half in oxide form. Sulfide copper extraction is about 50 percent and is made by leaching with ferric sulfate. Acid consumption is about 67 lbs/ton, which amounts to some 4.8 pounds per pound of copper recovered.

Pregnant solutions containing about 20 gpl of copper are stored in tanks and then go to precipitation launders filled with iron scrap. The solutions are forced to percolate and stripping is done in three stages, by which a 99.97 percent recovery is obtained. The average iron consumption is about 1.3 lbs per pound of copper precipitated. The

Fig. 5.6 Kennecott precipitation cones at Bingham mine. (Photograph by Don Green, courtesy of Kennecott Research Center.)

copper precipitate is removed from the launders by clamshell bucket and successively washed in a trommel to remove unconsumed iron. It is drained, dried to 14 percent moisture and shipped for smelting.

When the iron content of pregnant solutions is low, say below 10 gpl as it used to be at Chuquicamata prior to the introduction of the Exotica ore operation, instead of cementation, direct electrowinning of copper is possible. This is done at Inspiration, where a Dual Process was pioneered. As Dayton reported some years ago,[37] faced with the problems of mining lower-grade ores containing apart from oxides an increased proportion of sulfides, Inspiration abandoned the unique and classic acid-ferric sulfate leaching practice and switched to the Dual Process, which consists of straight acid leach of copper oxides and the flotation of washed residue to recover sulfide portions of the ore. The leached copper is then recovered by electrowinning. In some cases copper content of pregnant solutions is recovered by cementation with iron and then cement copper is dissolved for electrowinning.[38]

An ever-increasing quantity of copper is now being recovered by in situ leaching of old mines, dumps and even new but very low-grade orebodies. A classic example of an effective copper recovery operation from old mine dumps is at Bingham, where some 60,000 tpy of copper is recovered from a plant which was constructed at a total cost of about $20 million, i.e., about 8 to 10 times cheaper than the cost of a classic copper recovery installation of similar capacity. The second interesting example is Project Sloop by the same company.

The process applied by Kennecott at Bingham has the cone-type precipitators as a special feature, described in several papers.[53-54] In the precipitation cone, copper-bearing solutions are pumped through the manifold with the nozzles injecting them into a mass of iron. This is conducive not only to a rapid precipitation but to the removal of copper from the iron surface as well. In other words it is a self-cleaning device. The copper precipitates settle down through the stainless steel screen and accumulate on the sloped false bottom of the tank. Cones have proved to be a more effective means for precipitation than conventional launders because they produce a cleaner and more granular precipitate, with lower iron consumption as well. Typically, the purity of cement copper is about 90 to 95 percent and recoveries are also close to 90 percent and in some cases up to 95 percent.

Spedden, Malouf and Davis of Kennecott recently presented a very interesting paper on the theoretical aspects of in situ leaching.[55] This permits an understanding of the basic facts which underlie solution mining, when the grade of ore does not permit conventional mining and beneficiation of ores but nevertheless large underground masses of deposits or dumps contain appreciable metallurgical values. This study touches on an intriguing point in seeing solution mining as a process the reverse of mineral deposition and in seeking all possible acceleration of this process for the economic recovery of metals. For instance: utilizing the principle of secondary enrichment in copper porphyries, it is possible to envision an in-place leaching operation wherein the leaching solutions percolate downward through the interconnected thin fractures bringing about the dissolution of copper minerals. To speed up this dissolution, acid solutions and solutions containing ferric ions, oxygen and bacteria are used. The crucial questions for such in situ leaching operations are the permeability of the leached orebody, either natural or created by artificial fracturing such as in Project Sloop, by underground nuclear explosions.

As found by these researchers, dissolution of primary sulfides and the subsequent alteration to the secondary enrichment products of chalcocite and covellite lead to a molecular volume reduction that brings an increased permeability of the original site. The leaching of oxide copper values also increases the fracture void and opens new areas to solution flow.

Another important factor in leaching is the uniform wetting of the material to be leached. Malouf recommends moistening of waste material so that oxidation of sulfides can be promoted in the presence of atmospheric oxygen.[56] Atmospheric oxygen also diffuses through the slopes of the dumps up to 200 feet and is instrumental in the generation of heat due to oxidation of the sulfides. In general, when the rock is broken, the pyrite present in all copper porphyries tends to be transformed into sulfuric acid and ferric sulfate in the presence of humidity and oxygen. Iron is first oxidized to the ferrous state and then with additional oxygen into ferric ion. This oxidation can be accelerated by a number of bacteria: thiobacillus thiooxidans oxidizes sulfur while thiobacillus ferrooxidans oxidizes ferrous iron into ferric. The combined action results, then, in the formation of acid ferric sulfate, an excellent lixiviant for copper minerals.[57]

Malouf indicates [56] that the optimum pH for leaching is about 2. At pH 2.4 ferric iron salts are precipitated and tend to coat mineral surfaces and seal areas in the dumps, thus reducing leaching efficiency. Since these solutions should be constantly recycled to keep steams clean of pollution and to add to the leaching properties of solutions, acid is periodically added to keep pH about 2.1 or less.

Solvent Extraction

In cases where a relatively clean copper solution is wanted in order to be able to recover copper by electrolytical methods and where the presence of iron salts is not desirable, the solvent extraction method is rather helpful. It consists of stripping an aqueous acid solution of copper with an organic liquid, normally kerosene and a solvent, which permits the removal of iron and other contaminating salts. In this way copper values dissolved in a large volume of pregnant aqueous acid solution are transferred into a smaller volume of organic solution, normally consisting of kerosene and some additives like LIX-64 or others.

After the extraction cycle, the barren aqueous solutions (raffinate) are returned to the leaching recycle, while the pregnant organic solution advances to a scrubbing unit to remove any entrapped aqueous impurities. The copper values are stripped from the pregnant organic solution with an aqueous sulfuric acid solution to obtain a still smaller volume of strip solution which is the feed to the electrolytic copper plant. This last solution should contain from 10 to 25 grs/l of copper.

The basic prerequisites for such solvents are: selectivity for copper extraction and non-extraction of iron at natural low pH; low solubility of the solvent in water, to minimize solvent losses; nontoxicity for bacteria so they can continue to work after recycling; stability at temperatures between 0 and 80°C; low cost and low losses of solvent during the process.

This process is now increasingly used in complicated cases such as at Bagdad where very pure copper is being produced, or at Exotica, where leaching of copper results in leaching of many other salts, such as iron, aluminum and manganese. The most successful solvent reagents so far produced are LIX-63 and LIX-64, developed by General Mills,[58] which are respectively hydroxime and hydroxibenzophenoximes. LIX-63 was particularly useful for extracting copper from solutions produced by conventional ammoniacal leaching of oxide ores with alkali gangue. It has low solubility in water but cannot work in pH below 3, and thus is of little use for dump leaching solutions. On the contrary, LIX-64 is designed to work in acid media of pH between 1.4 and 3 and thus is ideal for the recovery of copper produced by any kind of leaching. It has a very high extractive power for copper, very small for ferric iron, molybdenum and vanadium, and none for other metallic ions. It should be used in diluted form with nine parts of kerosene, and in a typical operation has a loading power of 2.5 grs copper per liter of solvent mixture. Copper extraction is in the 95 percent plus range and raffinate contains as little as 0.03 grs/l of copper. This reagent has been used successfully at Bagdad, Esperanza and Inspiration and in spite of its relatively high cost of $2.50 per pound, the cost per pound of copper recovered, i.e., reagent losses, amounted to only 1.7 cents.[59-60]

Finally, quite recently General Mills announced the commercialization of a new liquid ion exchange reagent LIX-70.[61] This reagent was specifically designed to overcome some limitations of LIX-64. It was reported that LIX-64 was somewhat limited in its action by pH range. For instance when pH drops below 0.8 the extraction efficiency of LIX-64 drops. Thus in some copper streams where the copper content was rather low, below 5 grs/l, and the acid content very high — 25 to 50 grs/l — the efficiency of recovery of copper was rather low, in the 75 percent range. The new LIX-70 was designed to overcome these difficulties. It widens considerably the range of action of LIX-64 and while maintaining the same ultimate loading capacity can work with solutions of higher copper and/or acid content. It also rejects better iron giving a ratio of about 3000 : 1 in Cu : Fe. The future of LIX-70 is seen particularly bright for the extraction of copper from concentrates in so-called "smokeless" smelters, which should replace the conventional smelting and fire-refining techniques. By using LIX-70 in leaching of concentrates, cathode copper can be produced at mill-site.

Investment and Production Costs

A recent study by the staff of the U.S. Bureau of Mines[52] throws an interesting light on the close relationship between the copper price and the availability of copper supplies. The situation in the United States, which holds at least some 20 to 25 percent of the world copper reserves, is very representative for the whole Western World economy.

According to this study at a $0.50 per pound copper price (which has been predominant in recent years and very often exceeded in times of crisis and short supplies), considering a 12 percent rate of return on investment capital and credit for the recovery of by-products (see Chapter Four), the United States has some 83 million tons of economically recoverable copper. These resources will be accompanied by 1.8 million tons of recoverable molybdenum, some 43 million troy ounces of gold and 740 million ounces of silver. The average grade of recoverable copper ore will be, in this case, about 0.6 percent.

In the case of a $0.75 per pound copper price, domestic copper resources will grow to 115 million tons of metallic copper (a 0.55% Cu grade); and in the case of a $2 per pound copper, domestic resources in copper will grow to some 180 million tons of recoverable metal (average grade 0.46% Cu), plus 4.4 billion pounds of molybdenum, 60 million ounces of gold and 1,130 million ounces of silver.

The same report gives very interesting information on typical investment and operation costs as indicated in Tables 5.5 and 5.6.

TABLE 5.5

ESTIMATED CAPITAL REQUIREMENTS
FOR A FLOTATION CONCENTRATOR

(in 000 $ US of 1970)

	40,000 tpd	72,000 tpd
Crushing Plant - Primary	$4,152	$12,843
Crushing Plant - Secondary	12,268	17,149
Grinding Plant	29,139	34,920
Flotation Plant	15,004	13,406
Molybdenite By-Product Plant	1,526	12,579
Filter Plant	958	1,229
Lime Preparation Section	1,023	1,365
Tailings Disposal Section	4,909	10,359
Subtotal	68,980	103,851
General Facilities and Utilities	6,898	10,385
Subtotal for Depreciation, Taxes, Insurance	75,000	114,236
Working Capital	3,864	6,446
Total Capital Required	79,742	120,682
Per ton of Installed Daily Capacity	$ 2,000	$ 1,676

It will be observed in general terms that crushing plant and grinding plant installations amount to about 25 to 30 percent of plant costs and from 20 to 25 percent of the total capital investment; similarly, flotation plants are fully 35 to 42 percent of plant costs and 30 to 36 percent of total investment. By increasing plant capacity from 40,000

TABLE 5.6
ESTIMATED ANNUAL OPERATING COST FOR COPPER CONCENTRATORS

	40,000 tpd.			72,000 tpd.		
	Annual Cost	Total	Per ton	Annual Cost	Total	Per ton
Direct Cost:						
Power 34,700 KWH/hr x 8,568 hr/yr x $0.007/KWH	2,081,000			2,746,900		
Natural Gas 20 Mscf/hr x 8,568 hr/yr x $0.43/Mgcf	73,700			257,900		
Rods, balls, grinding & crushing surfaces - $0.149/ton	2,127,700			4,292,600		
Reagetns	1,116,700			2,270,600		
Other operating supplies (7.5% of power)	156,100	5,555,400	$0.39	206,000	9,774,000	$0.38
Labor 1,243 man hr/day x 357 days x $3.46/man hr.	1,535,400			2,203,600		
Supervision 15% of labor	230,300	1,765,700	$0.12	350,500	2,534,100	$0.10
Maintenance — supplies and parts	1,088,000			1,645,800		
Labor 835 man hr/day x 357 days x $3.65/man hr.	1,088,000			329,200		
Supervision 20% of labor	217,000	2,393,600	$0.17	1,645,800	3,620,800	$0.14
Payroll overhead		767,800	$0.05		1,127,300	$0.04
Total Direct Cost		10,482,500	$0.73		17,056,200	$0.66
Indirect Cost:						
Administrative, technical & clerical labor	364,000			588,000		
Payroll overhead	91,000			147,000		
Facilities maintenance & supplies (5% plant fac.)	362,600			553,400		
General overhead, office charges, research (5% direct cost)	524,000	1,341,700	$0.09	852,800	2,141,200	$0.08
Fixed cost:						
Taxes and insurance — 2% plant cost	1,595,400			2,435,100		
Depreciation — 15 years	5,058,500	6,653,900	$0.47	7,615,700	10,050,800	$0.39
Gross Operating Cost		18,478,100	$1.29		29,248,200	$1.14

tpd, the capital requirements per ton of installed daily capacity are reduced 16 percent, from $2,000 to $1,676 per ton. It is convenient at this point to observe how in spite of savings achieved through the building of larger plants, the costs per unit of installed capacity keep growing. In the late fifties and early sixties it was around $1,000 per ton of daily capacity, or somewhat below (Silver Bell, Pima, and Mission). In the early seventies it varies already from $1,700 (Sierrita) to over $2,000 (Tyrone and Lornex) and in the case of Twin Buttes will surpass $2,500.

Details in Table 5.6 indicate the operating costs of these plants in 1970: production costs can be reduced some 12 percent from $1.29 to $1.14 per ton of ore treated by increasing the capacity of the plant. The main savings are in labor, supervision and depreciation. At an average recoverable copper content of U.S. copper ores of 15 pounds per ton, the milling costs should be between 7.5 and 8.5 cents per pound of copper and in marginal cases of 8 to 10 pounds per ton of copper these costs can go as high as 16 cents per pound.

The same study indicates that a cementation plant using dump leaching and precipitation tanks and launders on a scale of 3,000 tpy of cement copper can produce copper at 5 cents per pound, while a solvent extraction and electrowinning plant with 7,000 tpy of electrolytic copper capacity and using LIX-64 produces copper at 10 cents a pound. The obvious advantages of these operations are the absence of mining costs and relatively simple equipment used in hydrometallurgical treatment, i.e., very low capital cost.

SELECTED BIBLIOGRAPHY FOR CHAPTER FIVE

1. Kennecott Copper Corporation Annual Reports, 1960 - 1972.

2. D. L. Simpson, B. H. Ensign, K. F. Marquardson: The Design of the Process for Copper Recovery from Silicate Ores at Ray Mines Division; AIME Meeting at Hayden, Arizona, May 10, 1968.

3. G. P. Sewell: Changes in Milling Practices at Hayden; AIME Meeting at Hayden, Arizona, May 10, 1968.

4. J. B. Huttl: The New Hayden Smelter Prepares to Take Ore; Engineering and Mining Journal, June 1959, pp. 97-104.

5. J. B. Huttl: The LPF Plant - Key to New Hayden Smelter; Engineering and Mining Journal, June 1959, pp. 104-107.

6. George Argall, Jr.: Taylor Metallurgy to Ore at Hayden; World Mining, June 1959, pp. 40-49.

7. Wayne H. Burt: Kennecott Expands Utah Copper Division; Journal of Metals, July 1966, pp. 819-823.
8. Paul G. Mahoney and Weston Starratt: Revamping largest copper smelter in the U.S.; Engineering and Mining Journal, April 1968, pp. 71-77.
9. W. A. Gibson and A. D. Trujillo: From Indian Scrapings: Development of Chino; Mining Engineering, January 1966, pp. 54-60.
10. Lewis Nordyke: Kennecott's Nevada Mines Division; The Explosives Engineer, May-June 1956, pp. 72-86.
11. A. W. Knoerr: San Manuel — America's Newest Large Copper Producer; Engineering and Mining Journal, April 1956, pp. 75-100.
12. V. B. Dale: Mining, Milling and Smelting at San Manuel; USBM Information Circular 8104, 1962.
13. The Significance of Sierrita; Metal Bulletin, London, October 23, 1970.
14. George P. Lutjen: Sierrita Makes it with Big Equipment; Engineering and Mining Journal, August 1970, pp. 70-73.
15. Thomas Jancic: How Engineering, Experience and High Capacity Equipment Combine to Make Ore at Sierrita; World Mining, March 1969, pp. 48-54.
16. Thomas Jancic: Development of Duval Corporation's Sierrita Mine; AIME Preprint 71-AO-63, New York, March 4, 1971.
17. R. W. Sayers et al.: Duval's New Copper Mines Show Complex Geology; Mining Engineering, March 1968, pp. 55-62.
18. Anthony Gomez, Jr.: Duval's Mineral Park Concentrator; AIME Meeting of Arizona Section, May 6, 1966.
19. C. H. Curtis: The Esperanza Concentrator; Mining Engineering, November 1961, pp. 1234-1239.
20. R. E. Thurmond et al.: Pima: A Three Part Story; Mining Engineering; April 1958, pp. 453-462.
21. M. D. Martin, G. A. Komadina and J. F. Olk: Pima Mining Company's Expansion; Mining Congress Journal, March 1966.
22. George A. Komadina: Two Stage Program Boosts Pima to 30,000 tpd; Mining Engineering, November 1967.
23. A. Blake Caldwell: Twin Buttes — A Deep Low-Grade Copper Producer; Mining Engineering, April 1970, pp. 51-66.
24. Twin Buttes Solves Problems of Great Depth, Low Grade, Complex Orebody; Engineering and Mining Journal, August 1970, pp. 74-78.

25. Ted Cook: Twin Buttes Sulphide Concentrator; AIME Arizona Section Meeting, 1973.
26. E. E. Krist: Copper Concentrate Spray Drying; AIME Arizona Section Meeting, 1973.
27. William Wrath: Anaconda's Butte Concentrator; Mining Engineering, May 1964, pp. 55-78.
28. John Huttl: New Success at Yerington; Engineering and Mining Journal, March 1962, pp. 74-85.
29. Another A for Anaconda; Engineering and Mining Journal, August 1954, pp. 71-94.
30. Yerington — Anaconda's Latest; Mining World, July 1954, pp. 45-64.
31. Norman Weiss and J. D. Vincent: Design and Operation of the Mission Mill; Mining Congress Journal: 3 parts, November, December, 1962, and January 1963.
32. ASARCO's Mission Copper; Mining World, January 1962, pp. 19-42.
33. J. D. Vincent and Wayne Bossard: Mission By-Products Plant Operation Control; AIME Meeting, February 1966, Preprint 66B70.
34. H. K. Martin: Milling Practice at the Lavander Pit Concentrator; Mining Engineering, November 1957, pp. 1229-1234.
35. John Huttl: Silver Bell — Uncle Sam Does it Again; Engineering and Mining Journal, July 1954, pp. 71-79.
36. W. R. Hardwick: Open Pit Mining and Concentrating at Silver Bell, ASARCO; USBM Information Circular 8153, 1963.
37. Stanley Dayton: Dual Process at Inspiration; Engineering and Mining Journal, September, 1957.
38. W. R. Hardwick: Mining and Costs at Inspiration; USBM Information Circular 8154, 1963.
39. John V. Beall: Southwestern Copper — A Position Survey; Mining Engineering, October 1965, pp. 77-92.
40. John V. Beall: Copper in the U.S. — A Position Survey; Mining Engineering, April 1973, pp. 35-47.
41. R. S. Shoemaker and A. D. Taylor: Mill Design in the Seventies; AIME Transactions, June 1972, pp. 131-136.
42. N. Weiss and J. H. Cheavens: Present Trends in Mill Design; Milling Methods in the Americas, VII International Mineral Processing Congress, New York, 1964.

43. Wada et al.: Some Experiments on the Kerosene Flotation; Studies from the Research Institute of Mineral Dressing and Metallurgy; Tohoku University, Vol. 15, Japan, 1968.

44. S. V. Dudenkov et al.: Theory and Practice of Flotation Reagents Use; Edition Nedra, Moscow, 1969.

45. R. W. Herlund: Extraction of Molybdenite from Copper Flotation Products; 50th Anniversary of Froth Flotation in the U.S.A.; Quarterly, Colorado School of Mines, Vol. 56, No. 3, 1961.

46. K. F. Beloglazov: The Froth Flotation Lows; Moskva, 1947.

47. Alexander Sutulov: LPF Process, University of Concepcion, 1963.

48. William Wraith, Jr. and T. G. Fulmor: Anaconda's Butte Concentrator; Mining Engineering, May 1964, pp. 55-78.

49. A. W. Last, J. L. Stevens and L. Eaton: LPF of semi-oxidized copper ores from Kennecott's Ray Mines Division; Annual AIME Meeting, New Orleans, February 1957.

50. P. Tilyard: Copper Cementation and its Application in LPF Process; Mineral Science and Engineering, Vol. 5, No. 3, July 1973.

51. J. L. Bolles: The Morenci LPF Process; 100th Annual Meeting of AIME, February 1971.

52. Harold J. Bennett et al.: An Economic Appraisal of the Supply of Copper from Primary Domestic Sources; U.S.B.M. Information Circular 8598, 1973.

53. H. R. Spedden, E. E. Malouf and J. D. Prater: Cone-Type Precipitators for Improved Copper Recovery; Mining Engineering, April 1966, pp. 57-62.

54. A. E. Back: Precipitation of Copper from Dilute Solutions Using Particulate Iron; Journal of Metals, May 1967, pp. 27-29.

55. H. R. Spedden, E. E. Malouf and J. Davis: In Situ Leaching of Copper; AIME Centennial Annual Meeting, March 1971, Preprint 71-AS-113.

56. E. E. Malouf: Current Copper Leaching Practices: Mining Engineering, August 1972, pp. 58-60.

57. Franklin D. Cooper: Copper Hydrometallurgy; U.S. Bureau of Mines Information Circular No. 8394, September 1968.

58. Copper Recovery from Acid Solutions Using Liquid Ion Exchange; Chemical Division, General Mills.

59. Solvent Extraction of Copper, Flow-sheet study Bulletin No. M7-F99; Deco Trefoil, November-December 1965.

60. J. S. Jacobi: The recovery of copper from dilute process streams; Mining Engineering, September 1963, pp. 56-62.

61. A major advance in liquid ion exchange technology — LIX-70; General Mills, AIME Centennial Annual Meeting, March 1971, 71-B-82.

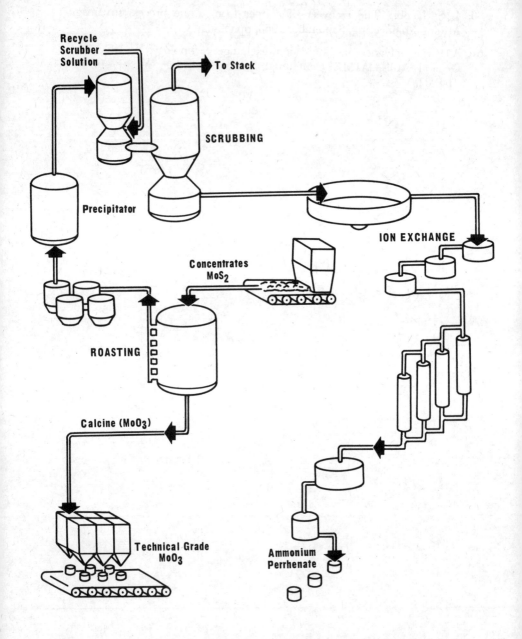

Fig. 0.6 A modern flowsheet for molybdenite processing and rhenium recovery. (Courtesy of Kennecott Research Center.)

CHAPTER SIX

Molybdenite and Rhenium By-Product Recovery*

In the previous chapter the principal points of primary molybdenite flotation were discussed and it was concluded that as a standard practice today, molybdenite (and thus rhenium) by-product is obtained from cleaner concentrates, normally the result of cleaner and recleaner flotation of primary rougher concentrates, as indicated in Fig. 5.2. The composition of these concentrates can be appreciated from the following reported operational data:

Composition of Selected Cleaner Concentrates

	% Cu	% Mo	% Fe	% S	% Insol
Mission	28	1.0	27	30	7
Silver Bell	30	0.45	24	NA	8
McGill	19	0.15	32	NA	10
Morenci	22	0.14	29	38	8
Copper Cities	26	0.28	27	32	12
Twin Buttes	29	1.1	28	NA	7
Pima	26	0.2	27	30	11
Sierrita	25	2.5	26	31	7
Chuquicamata	45	1.7	16	28	4
El Teniente	42	0.6	17	28	10
El Salvador	46	0.8	17	28	5

It can be stated then, that cleaner copper concentrates are at least 90 percent free from gangue minerals, that they contain between 50 and 80 percent copper minerals, a variable amount of pyrite and between 0.25 and 2 percent molybdenite, which should be separated from these concentrates. This striking disproportion between copper and iron mineralization on the one side and molybdenite content on the other, plus some problems which can be imposed by the character-

---

* This subject is extensively covered in a previous book by the same author: Molybdenum and Rhenium Recovery from Porphyry Coppers, 1970.

istics of gangue minerals, make very necessary a detailed study of the properties of molybdenite and its response to flotation phenomena before copper-molybdenum separation technology is discussed in any detail.

Molybdenite Occurrences and Properties

The case of by-product molybdenite recovery is a classical example of how a correct understanding of the mineralogical and geological conditions of a natural resource may bear clues for the solution of multiple metallurgical and production problems. Until a few years ago, most metallurgists paid relatively little attention to geological and mineralogical problems of molybdenum mineralization in an orebody and controlled their experiments and plant metallurgy entirely by chemical assays of heads, concentrates and tails. Such a system of "remote controls" is principally responsible for the existence of many theories, some of them correct, others wrong, about the peculiarities of molybdenite flotation, and there is no other way to bring some understanding to many strange phenomena but to start with a detailed description of molybdenite and its properties.

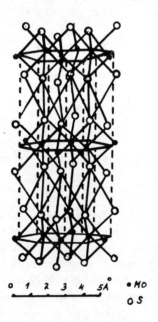

Fig. 6.1 Molybdenite crystalline structure.

Molybdenite is generally considered to be a very floatable mineral due to its laminar structure, which consists of two sheets of sulfur atoms between which, sandwich-like, lies a single sheet of molybdenum atoms bound with sulfur atoms by strong covalent bonds, as shown in Fig. 6.1. This distribution leads to the formation of natural laminae, in which the molybdenum atoms, all located in the same plane, are strongly held between two layers of sulfur atoms in S-Mo-S distribution. The bonds between laminae, however, are very weak, which explains a perfect cleavage along these surfaces. Since these surfaces consist exclusively of strongly hydrophobic sulfur atoms, the surfaces of the molybdenite laminae are also strongly hydrophobic.

This hydrophobicity does not apply to the edges of individual laminae where in the process of grinding the strong bonds between molybdenum and sulfur atoms are broken and where the exposed new surface is active with new reactions with other ions, water included. This fact gives rise to the so-called "edge effect," which explains a curious molybdenite characteristic, that with increased grinding, molybdenite hydrophobic properties decrease while the hydrophilic ones increase. As is clear now, with an increased degree of grinding, the proportion of edge surface increases as compared with the plate surface of molybdenite flakes, and this leads to the increased hydrophilic properties of molybdenite. Thus careful consideration to molybdenite grinding should always be given and overgrinding avoided.

Molybdenum, like the immediately lower and higher homologues, chromium and tungsten, from the 6-B sub-group of the Mendeleev Periodic System, forms a relatively small number of independent species. A little more than a dozen molybdenite minerals are recognized today in nature, molybdenite being by far the most important and abundant of them, and in fact the only mineral of economic interest. However, what help molybdenum minerals give to metallurgy in their relatively small diversity, molybdenite overweighs that with the abundance of its own problems. A detailed study of molybdenite in different copper porphyries has proved that the geological origin and mineralogical character of molybdenite in these deposits may vary from one orebody to another, and thus strongly influence flotation metallurgy.

Paragenesis, texture and dissemination of copper and molybdenum values, the degree of their oxidation, crystallization and mineralogical association with other minerals may be of great importance to the efficiency of the recovery and upgrading of these species in commercial products.[1] As the exploration studies of molybdenum bearing copper porphyries continue, more information on the geological and geochemical behavior of molybdenum is available and is very helpful for understanding many metallurgical problems encountered in complex copper-molybdenum flotation practice. Here, as in so many other cases, we can see with great clarity how metallurgical problems very often have their roots and unique explanation in geological and mineralogical phenomena.

In this context, for instance, it is important to observe that molybdenite mineralization in copper porphyries is generally uneven both chronologically and spatially. Molybdenite is normally deposited at a different geological time than the copper minerals. Its distribution generally does not follow the distribution of copper minerals, and at times appears to be erratic. No wonder, then, that the molybdenite concentrate output at different plants may vary considerably although no apparent change in copper mineralization and metallurgy has occurred.

Added to this distribution erraticity should be the uneven appearance of the molybdenite itself, which can appear in well formed crystals, with very fine dissemination, with or without a tiny oxide coat, associated or not with easily floated hydrophobic materials such as graphite, talc, pyrophyllite or carbonaceous materials.

To size up all these problems, very detailed studies must sometimes be made which take years to discover the nature of the problems. At El Teniente, for instance, for years it was a common belief that the orebody contained two types of molybdenite: one, very well crystallized, brilliant and floatable, and the other dull, amorphous and unfloatable. If the observation of this apparent difference was true, it could never be proven mineralogically. On the other hand, only relatively recently was it discovered that the so-called "dull" molybdenite was actually normal molybdenite but covered with an oxidized cap containing powellite.[2]

Similarly, at Chuquicamata where a serious oxidation problem is also observed, molybdenite flotation is helped by a less alkaline circuit between pH 9 and 10 instead of pH 11 or 12 at which copper flotation will be optimum.

Problems of Molybdenite Separation

As was already mentioned, out of several methods by which molybdenite can be recovered as a by-product from copper concentrates, the most popular and almost exclusively used is the one which contemplates the depression of copper and iron values and the flotation of molybdenite. This means that in all but two operations in the world the general flowsheet for molybdenite recovery consists of the following consecutive steps:

1. Thickening of the cleaner copper concentrate to eliminate an excess of reagents and to increase the concentration of newly introduced reagents.

2. Depression of copper and iron values by chemical means with or without previous heat treatment.

3. Flotation of rougher molybdenite concentrate under carefully controlled conditions of reagent addition, pH and percent solids.

4. Upgrading of rougher molybdenite concentrate by consecutive re-flotation and regrinding in a counter-current circuit.
5. Leaching, if necessary, of the final molybdenite concentrates to eliminate an excess of copper content.

Only in two operations, at Utah Copper and at Silver Bell, is the technology somewhat different, and essentially consists, first, in the depression of molybdenite with dextrine and the recovery of this product from tails. This is due to the fact that the other, easily floated gangue minerals like talc would follow molybdenite and then would be conveniently eliminated through a selective roast. When the tailings from copper flotation, which contain molybdenite and other gangue minerals like talc, are roasted at a temperature between 260°C and 320°C, then dextrine is destroyed and the surfaces of the molybdenite are passivized due to partial oxidation, while the floatability of talc and other gangue minerals is not affected. Thus the copper tail is thickened, filtered and subjected to a low temperature roast during 30 to 90 minutes. The hot calcine is repulped and, if necessary, given a wet grinding to break up agglomerations. Then, first talcose and other gangue minerals are floated. Once this troublesome material is eliminated, hydrocarbons and alcoholic frother are added and molybdenite concentrate recovered. This product is then upgraded in a standard way. Incidentally, Mission and Pima are also using a roasting step with multi-hearth roasters to eliminate talc, while Twin Buttes is equipped for this purpose with a spray drier, which will be described later.

Thermal Depression

It is now convenient to refer to Table 6.1 which represents a summary of all existing operating plants with the exception of Utah and Silver Bell, which have different processes, and Sierrita and Gibraltar which were not officially reported but which doubtlessly fall into the same category as the plants in Table 6.1.

It will be observed first that in all cases with El Teniente and Miami the only exceptions, prior to the concentrate retreatment, concentrates are thickened to 45 to 60 percent solids. This thickening is mainly done to eliminate an excess of reagents, which arrive with the copper concentrates from rougher, cleaner and recleaner circuits. Then, of course, this thickening is necessary along with filtering if some kind of roasting is intended.

In some operations, particularly in Russia, it has been found that heat treatment of copper concentrates in some cases may help

TABLE 6.1

MOLYBDENITE RECOVERY FROM COPPER CONCENTRATES

Plant	Thickening - % Solids	Heat Treatment: Steaming	Pressure Steaming	Kiln Roasting	Cooking	Copper Depression: Sodium Ferrocyanide	Sodium Cyanide	Nokes - P_2S_5 + NaOH	Nokes - As_2O_3 + Na_2S	Sodium Sulfide	Sodium Hydro Sulfide	Others	Molybdenite Flotation: pH	Percent solids	Fuel Oil	Kerosene	Pine Oil	MIBC	Others	Sodium Silicate	Number of flot. steps	Leaching of copper
San Manuel	55	–	–	–	–	X	–	–	–	–	–	X	8.0	30	X	–	–	–	X	–	6	–
Pima	50	–	–	–	–	–	C	–	–	X	X	–	11.0	35	X	–	–	–	X	–	9	–
Twin Buttes	50	–	–	–	–	–	–	–	–	X	–	–	10.5	35	X	–	–	–	X	–	8	–
Ray	60	X	–	–	–	X	C	X	–	–	–	X	8.5	20	X	–	–	X	–	–	9	–
Chino	50	X	–	X	–	–	–	C	–	X	–	–	–	15	X	–	–	X	–	–	12	–
McGill	60	X	–	–	–	X	C	–	–	–	–	–	9.0	30	X	–	–	C	C	–	8	–
Inspiration	60	–	X	–	–	–	–	X	–	–	–	–	9.5	35	X	–	–	X	–	–	7	X
Mission	60	–	–	X	–	–	–	X	–	X	–	X	8.5	30	X	–	X	–	–	X	10	–
Mineral Park	50	–	–	–	–	X	–	–	–	–	–	–	2.7	20	X	–	–	X	–	–	8	–
Esperanza	50	–	X	–	–	X	–	–	–	–	–	–	7.0	20	X	–	–	X	–	–	8	X
Miami	No	–	–	–	–	–	C	X	–	–	–	X	12.0	15	X	–	X	–	X	–	8	–
Bagdad	50	–	–	–	X	–	C	X	–	–	–	–	8.5	20	X	–	–	X	–	X	12	–
Chuquicamata	45	–	–	–	–	–	C	–	X	–	–	–	11.0	35	–	–	–	–	–	–	8	X
El Teniente	No	–	–	–	–	–	C	X	–	–	–	–	8.5	20	–	X	–	–	–	X	9	–
El Salvador	60	–	–	–	–	–	C	–	X	–	–	–	11.0	45	X	–	X	–	–	–	7	X
Toquepala	60	–	–	–	–	X	C	–	–	–	–	X	8.0	30	X	–	X	X	–	–	9	X
Lornex	50	–	–	–	–	–	C	–	X	–	–	–	8.5	35	X	–	X	–	X	–	7	–
Island Copper	65	–	–	–	–	–	C	–	–	X	X	–	10.5	35	X	–	–	–	X	–	8	–
Brenda	65	–	–	–	–	–	C	–	–	X	–	–	8.0	25	X	–	–	X	–	–	12	X
Gaspe	50	X	–	–	–	–	X	–	–	–	–	X	8.0	25	X	–	X	–	–	–	14	–
Balkhash	45	X	–	–	–	–	–	–	X	–	–	–	11.0	45	–	X	–	–	–	X	7	X
Almalyk	60	X	–	–	–	–	–	–	X	–	–	–	11.0	25	X	X	–	–	–	X	6	X
Medet	50	X	–	–	–	–	–	–	X	–	–	–	8.0	20	X	–	–	–	–	X	6	X
Kadzharan	60	X	–	–	–	–	–	–	X	–	–	–	10.5	25	–	X	X	–	–	X	6	X
Agarak	50	X	–	–	–	–	–	–	X	–	–	–	10.5	25	–	X	X	–	–	X	6	X
Sipalay	55	–	–	–	–	X	X	–	–	–	–	X	8.0	35	–	X	–	–	X	X	10	–

C - Cleaner Circuit

efficiency in copper depression which happens mainly due to the efficient destruction of residual flotation reagents. In many cases, as in those of the Kennecott properties at Ray, Chino and McGill, at Gaspe and in all Communist Bloc plants, steaming at about 100°C will be sufficient to destroy residual flotation reagents. This is carried out either by live steam introduced into the pulp under pressure, or by spiral heat exchangers. All depends on the temperature desired, its control and the agitation required. At Inspiration and Esperanza, steaming is done under pressure because of specific metallurgical problems. In both cases autoclaves are used.

If steaming does not provide enough heat for the effective destruction of flotation reagents (or dextrine when reactivation of moly is required), then skin roasting, generally in multi-hearth furnaces, is used. This is a low temperature operation which is carried out at about 300°C and which should be carefully controlled to avoid excessive oxidation of molybdenite surfaces while effecting oxidation of the copper sulfide particles. It is best controlled by the acidity of the coke. At Chino roasting is used as a complementary means to steaming, while at Mission it is intercalated into a molybdenite upgrading process: after elimination of copper by sodium sulfide and hydrosulfide depressors, the rougher concentrate is thickened, filtered and roasted for talcose material flotation, then conditioned with fuel oil and pine oil and refloated 8 times to obtain a high-grade product.

At Bagdad Copper, heat treatment takes place in the form of cooking concentrates prior to molybdenite recovery. They are thickened, conditioned with sulfuric acid to pH 5.5 and then cooked for about 1 hour just below the boiling point. Under these conditions, complete destruction of xanthates used in primary rougher flotation is insured.

In Russian plants it has been found that live steam treatment of copper concentrates drastically reduces the consumption of sodium sulfide.[3] It is postulated that at higher temperatures oxygen concentration in pulp is considerably reduced, which automatically decreases the rate of decomposition of sodium sulfide and also helps the desorption of flotation reagents. N. E. Plaksa, for instance, reports that at Balkhash, by introducing live steam heating, sodium sulfide consumption was reduced from 42 to 5.5 pounds per ton of concentrate, while maintaining the same recoveries.

In some cases this heating action is not sufficient to destroy residual reagents and it is helped by chemical means, as at San Manuel, with hydrogen peroxide and sodium hypochloride. This technique, however, has not found very wide application because of corrosion problems and damage to the equipment.

Copper Depression by Chemical Means

Thermal depression is always considered an auxiliary operation whose purpose is to reduce the costs of chemical depression of copper minerals. This depression was studied in considerable detail by various companies and research centers, and various patents exist to that effect. Among the most popular means for copper depression is the use of cyanides, including ferrocyanide and ferricyanide; then the so-called Nokes reagents, which are reaction products between phosphorus pentasulfide and sodium hydroxide, or between arsenic trioxide and sodium sulfide, known under the commercial name of ANAMOL. Then there is the very effective sodium sulfide depressor extensively studied and applied in Russia: sodium hydrosulfide, ammonium sulfide and various other similar reagents.

The use of cyanide and particularly of ferrocyanide for copper depression is based on an observation of Gaudin that ferrocyanide is a good depressor for chalcocite. This gave a basis to Barker and Young for their research and U.S. Patent No. 2,664,199, known under the name of the Morenci Process.

According to this process, copper and iron sulfides can be efficiently depressed by sodium ferrocyanide into a pulp of 20 percent solids and a pH 7.5 to 8.5. The mechanism of reaction seems to be rapid desorption of xanthates from the chalcocite surface. However, the depressing action wears out with time apparently because cyanide reacts with xanthates and copper ions at the chalcocite surface to form soluble copper complexes. When all free cyanide has reacted, xanthate anion can resorb again.[4] Thus adding cyanides in stages is recommended. The controlling factors of the process seem to be pulp density, pH and cyanide concentration. The final stages of purification of molybdenite concentrates are made with pH between 9 and 11.5 in a countercurrent system and with the help of sodium cyanide.

Golikov and Nagirniak[5] have studied the depression of chalcopyrite and pyrite with cyanides when floating these minerals with different xanthates and dixanthogens. They have found that when floated with lower xanthates, pyrite depresses more easily than chalcopyrite, but the opposite is true when flotation is made with xanthates above butyl xanthate. In general, depression with cyanides is more difficult if the radicals of xanthates are longer and if the minerals were in contact with collectors before flotation. The influence of dixanthogenes is not pronounced, and generally less cyanide is necessary for the depression of sulfide floated with these reagents.

The Morenci Process has had wide application, particularly in operations with heavy chalcocite mineralization. The process needs adjustments and is experiencing modification at Morenci itself. It is presently used with some modification in 25% of copper porphyries.

Nokes Reagent

Another very effective means to depress copper and iron sulfides is by using Nokes reagent, which is used today in many large copper operations recovering molybdenite by-product, in North and South America.

This instantaneous depressor, which is a product of the reaction of stechiometrical quantities of sodium hydroxide and phosphorus pentasulfide, was developed by Nokes during the Second World War, between 1943 and 1944, in order to replace the somewhat tiresome roasting treatment practiced at Kennecott properties. Its objective was to replace heat treatment of concentrates with chemical treatment and to simplify molybdenite by-product recovery.

The merits of this extremely efficient reagent were not clearly understood by industry for many years. However, with time its popularity has increased, and now 10 operating plants use it in one form or another.

This reagent was discovered by Kennecott research workers Nokes, Quigley and Pring, and was registered under U.S. Patent 2,492,936 in 1949, 2,811,255 in 1957 and Chilean Patent 10,264 in 1946.

The patent claims that polysulfides based on compounds of phosphorus, arsenic or antimony as carriers of sulfur and hydroxides of sodium, calcium, ammonium, potassium and others can be formed by exothermic reactions when sulfides are mixed in a water medium with hydroxides. The essential characteristics of the reagent, as claimed by the inventors, are: 1) the presence of bivalent sulfur, 2) the presence of oxigene and 3) the participation of either phosphorus, arsenic or antimony in the compounds.

In fact, no one knows clearly what kind of complex compound is formed in this reaction. Phosphorus pentasulfide is already by itself a very complex compound with a global formula of P_2S_5. It is likely that the exothermic reaction which takes place when this reagent is put into contact with an aqueous solution of sodium hydroxide produces not one but several complex products.

Kennecott generally uses the product of the reaction between phosphorus pentasulfide and sodium hydroxide which it calls LR-744 (LR for Laboratory Reagent). Anaconda has produced its own reagent reacting sodium sulfide with arsenic threoxide, and calls it ANAMOL D (Ana for Anaconda, and Mol for molybdenum).

At one time it was thought that the action of Nokes reagent was based on detergent properties and that what the Nokes reagent actually did was desorb anionic collectors from copper and iron sulfides.

This theory apparently has little basis in fact because the copper or iron sulfide ores treated with the Nokes reagent remain lastingly depressed and are hard to float again with any other reagent.

As indicated, the reagent has an instantaneous action and thus should be added immediately to flotation. Also it reacts very quickly with mineral particle surfaces and is rapidly consumed, for which reason stage addition is preferable. The normal consumption varies greatly with the nature of the ore and may be between 4 and 10 pounds per ton of concentrate. The efficiency of depression is optimum at pH 8.5 and higher pulp density gives better results than diluted pulps.

The use of Nokes reagent produces at times problems in froth control which occasionally turn uncontrollable. Previously repulping with fresh water was a means to counteract this effect, while today a number of antifoam reagents are available for this purpose.

The Nokes reagent apparently works with all kinds of copper mineralizations, starting with primary chalcopyrite and bornite, through the mixed chalcocite, chalcopyrite and pyrite ores, and with secondary chalcocite ores. However, its efficiency was particularly proved with copper minerals containing iron.

The classical application of the Nokes reagent does not require any previous heat treatment. The cleaner concentrates should be conditioned with this reagent at higher percent solids, and rougher molybdenite concentrate can be recovered with a hydrocarbon collector and an alcoholic frother in an alkaline circuit which may vary from 8.5 to 11.5. This is the way molybdenite is recovered at Inspiration, Miami, El Teniente, Chuquicamata and El Salvador. In the cases of Miami and El Teniente, this treatment is done even without previous thickening.

In other cases the action of the Nokes reagent is helped by previous steaming of the pulp. Such is the case of Bagdad Copper, and this was used previously at McGill. Finally, in some cases like Chino, Ray and Mission, Nokes is an auxiliary depressor used in recleaner circuits to purify and decopperize molybdenite concentrates.

Generally speaking, the Nokes reagent, because of its high consumption, may be an expensive depressor. Thus its use on a minimum volume of pulp is preferable. It can produce very good basic copper-molybdenum separation, and in combination with sodium cyanide can upgrade molybdenite concentrates up to maximum specifications in a counter-current system. Intercalation of flotation and regrind operations proves to be useful for opening and "polishing" new surfaces.

The Russian Sulfidization Process

The use of sodium sulfide for the depression of copper and iron sulfides in molybdenite flotation was discovered in the Soviet Union and patented by Vartanian and Gomelauri as Russian Patent No. 48,010 in 1936 and No. 63,803 in 1941. Later, the problem was theoretically studied by Mitrofanov, Kurochkina, Sokolova and many other metallurgists, which has resulted in a great amount of experimental work and publications. This is probably one of the best theoretically studied processes for molybdenum recovery while depressing copper and iron values.

At the present time, this process is used in all Russian plants, notably at the Balkhash, Almalyk, Kadzharan, Agarak, Dastakert and Sorski plants, at Rosen and Medet in Bulgaria, and, since 1966, at Mission in Arizona. Mission has obviously used Russian experience in this matter, and is the first known case of using Russian technology in the West. Altogether, this type of sulfidization process accounts for 20 percent of the cases and tonnage treated.

Wark and Cox have observed that the adsorption of a collector on the surface of sulfides depends critically on the concentration of sodium sulfide in solution. In other words, a maximum concentration of sodium sulfide exists which still permits the adsorption of collectors on the surface of minerals. If this concentration is exceeded, the collectors are not only adsorbed but may as well be desorbed. In this way sodium sulfide, if in sufficiently high concentrations, may be used for the depression of most sulfides.

The depression of galena, sphalerite, pyrite, chalcopyrite, chalcocite and many other copper and non-ferrous sulfides by sodium sulfide has been studied and proved by many scientists. Among them should be mentioned Konev, Debrivnaia, Glazunov, Wark, Cox, Gaudin, Schwedov, Kremer, Glembocky, Mitrofanov and others.

Glazunov and Ratnikova have proved[4] that the adsorption of sulfide ions on mineral particles and thus their depression depends on pH.

This certainly may involve different mechanisms of depression, since at one pH the solutions of sodium sulfide may have a maximum concentration of hydrosulfide ions and at another, of sulfide ions and hydrogen sulfide.

The highest velocities of adsorption of sulfides were observed on the surfaces of galena, followed in decreasing order by pyrite, sphalerite and chalcopyrite.

Mitrofanov and Kurochkina have proved that adsorption of sulfide ions on molybdenite surfaces is considerably lower than on the surfaces of chalcocite. Moreover, with an increase of pH, while the adsorption on molybdenite decreases, the adsorption of sulfides on chalcocite increases. The experimental data are approximately as follows:

	Adsorption mgs/l	
pH	*Chalcocite*	*Molybdenite*
3	350	1.6
5	380	1.6
7	480	1.7
9	520	1.1
11	620	0.7
12	700	0.5

Taking into consideration the absolute values of this adsorption, it is clear that while different sulfides are affected by sodium sulfide, molybdenite is practically not. This permits molybdenite flotation with great selectivity. The selectivity (and lesser adsorption of sulfide ions) is also improved if the molybdenite is previously conditioned with hydrocarbon collectors.

In their patent, Vartanian and Gomelauri observe that in order to obtain a satisfactory depression of sulfide, a concentration of free sodium sulfide of 200 mgs/l should be maintained. The control can be automatically carried out with potentiometers.

The consumption of sodium sulfide, which is generally high and varies according to the situation from 10 to 80 pounds per ton of concentrate, depends on three basic factors: 1) the mineral composition of the concentrates; 2) the flotation time of molybdenite; and 3) the pH of the circuit.

In general, sodium sulfide consumption increases with the increase of the content of oxidized copper minerals in the concentrate. It is also thought that the surfaces of copper minerals, pyrite and galena catalyze the oxidation of bivalent sulfur to hexavalent sulfur, thus causing the reagent consumption increase.

On the other hand, flotation time also influences reagent consumption because with time the depressing effect of sodium sulfide wears out and fresh amounts of reagent are necessary. This explains the stage addition of reagents. Oxygen dissolved in water is also an oxidant for Na_2S. Finally, at low pH — in acid conditions — sodium sulfide tends to hydrolyze, producing H_2S which increases Na_2S consumption. At higher pH, this hydrolysis is reduced and hence a decrease in Na_2S consumption occurs. Thus, depression should be carried out in the alkaline circuit.

Dudenkov has demonstrated that the consumption of sodium sulfide can be decreased in cleaner circuits of some phosphates, particularly when three sodium phosphate is added.

It is fundamental that the sodium sulfide used has a certain technical purity and not contain carbon or coke, since this material will float with molybdenite and contaminate the concentrate. Besides, by smearing, carbon can activate the flotation of quartz and other gangue minerals.

Western Sulfidization Processes

As mentioned before, Mission Copper uses the Russian sulfidization process in basic copper-molybdenum separation.* The cleaner copper concentrate is thickened to 60% solids, conditioned with 7.2 pounds per ton of sodium sulfide, and floated in a rougher molybdenite flotation circuit. The rougher molybdenite concentrate is then dewatered, roasted, reground, repulped and refloated with the use of Nokes reagent for further copper depression. The economics apparently work better this way because sodium sulfide is cheaper than the Nokes reagent. Also, the apparently low consumption of sodium sulfide at Mission is due to the good automatic control of the circuits, the high percentage of solids in conditioning and the good quality of the depressant.

At Pima, which since 1967 has been in the process of developing its own depressing process, depression of copper minerals is done after thickening to 60% solids for the exclusion of flotation reagents and the dilution of the pulp to 54% solids. As a primary sulfidizer, a mixture of 85% sodium hydrosulfide and 15% ammonium sulfide is used in

---

* Vincent and Bossard mention in their paper[f] that sodium sulfide and hydrosulfide were first used in 1933 by Anaconda at Cananea, Mexico.

the form of a 40% solution. The reagent consumption is about 16 pounds per ton and flotation is carried out at pH 11, with the percent of solids between 30 and 40 percent.

The depressing action of this sulfidizer is fortified with a zinc cyanide complex which is prepared from two parts of sodium cyanide and one part of zinc sulfate which is used in the cleaner stages.

No attempt is made to obtain a very high grade of the final product, by which recovery is considerably increased. As in the Russian practice, molybdenite activation with fuel oil is recommended prior to the application of depressors, otherwise molybdenite recovery may be somewhat affected. Another use of ammonium sulfide as a copper depressor is reported at Miami with its new flowsheet. Here the primary separation is made with the Nokes reagent, but ammonium sulfide is used for depression in cleaner circuits. In fact, the efficient use of ammonium sulfide apparently requires a high percent of solids in conditioning which takes between 30 and 60 minutes. Since no thickening facilities for the cleaner copper concentrate exist at Miami, ammonium sulfide has been replaced by Nokes in rougher molybdenite separation.[7]

Another kind of sulfide depressor has been created at Inspiration. Here a unique problem of sulfur flotation into molybdenite concentrate should be faced as the result of leaching operations, during which chalcocite in contact with ferric ions is transformed into artificial covellite with free sulfur liberation. In the beginning it was thought that sulfur may be eliminated by the roasting operation, but it proved to be a difficult task since the melted sulfur tended to solidify in roaster beds and brake the roaster arms.

This has justified, then, the application of autoclaves for the elimination of sulfur. Lime was added to react with sulfur in autoclaves and calcium polysulfide was formed, which proved to be a depressant for copper minerals. This certainly did not eliminate the use of the Nokes reagent for copper depression but has reduced the consumption.

Molybdenite Flotation

Since the molybdenite content of the copper cleaner concentrates is rather low, molybdenite flotation is carried out at a relatively low percent of solids, which generally fluctuates between 25 and 35 percent. Only in exceptional cases and with severe water problems can it go up to 45 percent solids.

Flotation of molybdenite is carried out generally with fresh water because of the sensitivity of metallurgy to foraneous ions. A strict pH control is generally necessary and carried out automatically for optimum metallurgical results. The alkalinity of the molybdenite circuit may vary within a wide range from pH 7.5 up to 11 depending on the process of copper depression selected. When using ferrocyanides pH is generally low, between 7 and 8, and must be strictly controlled. At Mineral Park,[8] the copper-molybdenum concentrate is steamed and heated in a pressure vessel, then cooled, conditioned with sodium ferrocyanide and floated in a rougher circuit at 20% solids. The rougher concentrate is thickened, reground and refloated in a cell-to-cell recleaning operation. On the other hand, at San Manuel the thickened cleaner concentrate is first attacked with sodium zinc cyanide and hydrogen peroxide, under strictly controlled pH conditions in both cases, and only then is copper depressed with ferrocyanide at pH 7 and molybdenite floated at 30 percent solids.

Nokes depressors also work best at low pH ranges, generally at pH 8.5, which is incidentally the optimum for molybdenite flotation. Such is the case at Bagdad, Ray, Mission and El Teniente. However, at Inspiration and particularly at Miami where serious problems with iron exist, higher pH is used. The same is the case with ANAMOL, used at Chuquicamata and El Salvador, where copper metallurgy requires a relatively high pH, 11, although it affects negatively molybdenite flotation.

It will be observed from Table 6.1 that in the great majority of cases fuel oil is used as a promoter for molybdenite flotation. The light fuel oil is preferred over the heavier diesel or stove oil because it contains more favorable components and does not affect the froth. However, in all Communist Bloc countries and at El Teniente, kerosene is used instead of fuel oil. The increasingly favored frother in the West is MIBC, though some other alcoholic frothers are also available. In some places pine oil is still used because of tradition and cheapness. In some western operations and particularly in Russia, sodium silicate is added to the circuits for the dispersion of slimes. This is particularly important when sericitic material is present.

Among the serious problems encountered in molybdenite upgrading is the elimination of the readily floatable non-metallic components of porphyry ores which tend to float along with molybdenite. The problem of talc, sericite and other gangue minerals has already been discussed. The solution to this problem was found in the roasting of concentrates and differential flotation of gangue prior to molybdenite. Also, in 1971

Kennecott reported [9] a new acid bake-leach flotation process, which is concerned with the upgrading of impure molybdenite concentrates. By heating impure concentrates at 260° to 290°C for one hour with concentrated sulfuric acid, contaminating sulfides are leached and gangue minerals activated for flotation. In this way concentrates containing up to 30 percent of insoluble and up to 10 percent of combined copper and iron and 3 percent lead could be purified to a product of 88.7 percent MoS_2, 0.34 percent copper and less than 0.02 percent lead.

The other readily floatable material which often contaminates molybdenite concentrate and is difficult to get rid of is graphite and carbonaceous gangue. It was first seriously observed in Russia at the Tuimski plant in Siberia,[10] where sodium sulfide used for the depression of copper values was contaminated with residual carbon due to the method of its fabrication. This carbonaceous material not only floated itself, but activated by smearing the other gangue particles as well, and made it almost impossible to obtain molybdenite concentrate of high purity. Later similar problems were observed at Morenci and Island Copper. At Morenci coal contamination arrived from the sponge iron which was added to the LPF circuit, while at Island Copper the carbonaceous material apparently came with the ore.

To eliminate all impurities to a reasonable extent and to separate molybdenite from copper and iron sulfides, normally from 8 to 12 flotation steps are necessary. They are normally carried out in a sequence of rougher flotation followed by the regrinding of rougher concentrate and the counter-current flotation of cleaner and recleaner concentrates. Reagents are added as necessary and the action of main depressors is often fortified by sodium cyanide in the last stages of purification. However, even the most persistent retreatment quite often cannot attain the desired purity of molybdenite concentrates. Apart from this, the attempt to obtain very pure concentrates is accompanied by heavy losses in recovery. Plant production can be reduced by 20 and 30 percent due to this factor alone.

In Russia where the molybdenite content of copper concentrates is extremely poor and where great losses have been experienced during the attempts to achieve a high-grade of products, this subject was approached in a different way. Molybdenite is concentrated only to a certain purity of, say, 60 percent, or 36% Mo, and then treated by roasting and hydrometallurgy. This would allow not only the obtaining of quite pure products such as molybdenum threeoxide, but the recovery of numerous by-products, rhenium among them.

In the West this approach also has had some response in difficult metallurgical cases. At Pima, for instance, where a hydrosulfide-ammonium sulfide depressor is used for basic depression and sodium cyanide for cleaner circuits, in order to preserve recovery a rather unclean product is produced whose assays are as follows:

Mo	Cu	Fe	S	Insol
38.3%	2.07%	4.4%	30.9%	18.1%

The point is to keep copper under 2.5% which permits obtaining a price which is still favorable. If the concentrate is contaminated heavily with talc, the insoluble can be removed after a partial roast.

Molybdenite concentrates are subject to very stiff impurity specifications because most of them go into the steel industry for alloys fabrication. They must be very low in copper, lead, arsenic and phosphorus. These specifications are normally easily met with the exception of copper, since too much retreatment causes heavy metallurgical losses. Thus a common practice today is to adjust the final copper content of concentrates by leaching The most popular leaching medium is sodium cyanide which easily dissolves chalcocite, covellite and some other secondary copper minerals. It is presently used in almost a third of the plants operating. For instance, at Inspiration where a 90% MoS_2 concentrate with 2 percent copper is obtained by flotation retreatment, leaching lowers the copper content to less than 0.5% Cu, and at Chuquicamata and El Salvador copper content after leaching is about 0.2%. At Esperanza, however, the leaching of copper is effected after molybdenite concentrate roasting, which also results in a very pure product.

However, in some cases copper contamination is chalcopyritic, and chalcopyrite would not leach with cyanides. At Brenda this requires a leaching with hot ferric chloride which reduces the copper content to 0.1 percent. Other impurities such as arsenic (when ANAMOL is used for depression) or alkali salts can be removed by repulping if specifications are not met.

Rhenium in Porphyry Coppers

As has already been indicated several times, the only known commercial source for rhenium is the molybdenite concentrates recovered from porphyry coppers. The other known sources are too low in rhenium content to justify recovery.

Table 6.2 gives a summary of the rhenium content in some major copper porphyries, expressed in grs/metric ton or in parts per million

Fig. 6.2 Kennecott's molybdenite roasting and rhenium recovery plant at Magna, Utah. (Photograph by Don Green, courtesy of Kennecott Research Center.)

on 100% MoS$_2$ concentrate. The copper porphyries of the United States contain an overall average of about 0.04 percent rhenium, or 400 ppm or 0.8 pounds per ton. The highest rhenium content is found at McGill, a Kennecott Copper property, which has over 3 lbs of rhenium per ton of molybdenite. Taking into consideration that rhenium market prices vary between $1,000 and $1,200 per pound, the value of the rhenium in such a concentrate exceeds the value of the molybdenite itself. A very high concentration of rhenium is also found in San Manuel molybdenite concentrates. In general, the Arizonian copper porphyries are rather high in rhenium content, between 500 and 600 ppm, with the exception of Sierrita which is pretty low in rhenium in the same way as Brenda Copper Mines in British Columbia. Both mines have a very low copper and very high molybdenum content. In British Columbia at Island Copper the highest rhenium-bearing molybdenite concentrate is produced. It contains 4 pounds of rhenium per ton of molybdenite, and thus should be considered a premium material. This is particularly important now when apparently an excessive supply of molybdenite exists on the market.

Latin American copper porphyries are considered to contain an average of 400 ppm of rhenium. Molybdenite concentrates from El Salvador, El Teniente and Cananea are rather high — in the 500 to 700 ppm range, while those from Andina, La Disputada, Toquepala and Chuquicamata are in the 300 ppm range. The Argentinian porphyry coppers are not very high in rhenium. Some samples indicate a content up to 150 to 200 ppm.

The Russians also have quite a wide variety of molybdenite concentrates. Those from Armenia are not very rich — in the 300 ppm range, while the Kazakhstan porphyries are in the 500 ppm range. Almalyk in Uzbeckstan contains only 290 ppm.

In Table 6.3 an attempt has been made to assemble all available statistical data to evaluate the world situation in rhenium resources, actual and potential production.

It will be observed that generally the estimated 5.5 million tons of world molybdenum reserves in porphyry coppers contain about 4.2 million pounds of rhenium, i.e., an average of 0.76 pounds of rhenium per ton of molybdenite. If we take into consideration that present technology permits a 50 to 60 percent molybdenite recovery from copper concentrates and, say, a 60 percent recovery of rhenium from molybdenite concentrates, then from this total only 1,250,000 to 1,500,000 pounds of rhenium can be considered a recoverable product.

TABLE 6.2

RHENIUM CONTENT OF PORPHYRY COPPERS

(expressed in ppm on 100% MoS_2)

North America		South America	
McGill	1,600	Chuquicamata	230
San Manuel	1,000	El Teniente	440
Chino	800	El Salvador	570
Cities Service	600	Andina	380
Twin Buttes	600	La Disputada	350
Pima	600	Toquepala	325
Mission	600	Argentinian porph.	170
Bagdad	200	*Communist World*	
Esperanza	200	Kounrad	510
Sierrita	180	Almalyk	290
Mineral Park	60	Kadzharan	300
Island Copper	2,000	Aigedzor	1,000
Brenda	80	Dastakert	80
Cananea	700	Medet	125

However, at the present time only 35,000 tons of molybdenite containing some 37,500 pounds of rhenium are being recovered from porphyry coppers per year. Since the market for rhenium is still very limited — mostly because of the insecurity of supplies — only some molybdenite concentrates, particularly those with a high rhenium content, are being processed for rhenium recovery. The largest rhenium producer in the world is the United States with an estimated output of 6,000 pounds per year, based mainly on Kennecott production in Utah and in Cleveland. The second world producer is considered to be the Soviet Union, which established its production at Balkhash in 1948. The estimated output from primary sources is about 1,400 pounds per year, but this figure may be considerably higher if some other additional sources are considered. The Russians have done very extensive research in rhenium chemistry and metallurgy and apparently possess quite an advanced technology.

Canada will very soon be a very prominent rhenium producer or exporter due to its unique high-grade orebody at Island Copper. Although at this writing no specific arrangements for rhenium recovery from molybdenite are known, they must be coming very soon simply

TABLE 6.3

RHENIUM RESOURCES AND PRODUCTION FROM PORPHYRY COPPERS

Areas	Resources			Production		
	Molybdenite Reserves - tons	Average Re Content Percent *	Rhenium Content - pounds	Molybdenite Production - tons	Potential Re Production - pounds	Actual Production - pounds
United States	1,700,000	0.04	1,360,000	17,000	22,200	5,900
Canada	350,000	0.05	350,000	6,800	6,800	2,000**
Chile	2,150,000	0.04	1,720,000	7,000	5,600	1,500***
Latin America	850,000	0.03	510,000	750	225	–
Soviet Union	180,000	0.04	144,000	3,000	2,400	1,400
Others	270,000	0.03	162,000	450	250	–
	5,500,000	0.038	4,246,000	35,000	37,500	10,800

* Calculated on 100% MoS$_2$
** Not acutally recovered, but will be very soon
*** Recovered from Chilean molybdenite concentrates, but not necessarily in Chile.

Fig. 6.3 Scrubbing system for gases from molybdenite roasting plant in Magna. (Photograph by Don Green, courtesy of Kennecott Research Center.)

because with the present price of rhenium the rhenium content of molybdenite concentrates exceeds the value of the molybdenite itself.

Chilean rhenium production potentialities can be clearly seen from Table 6.4. Presently, Chile produces about 13 million pounds of molybdenum (about 24 million pounds of molybdenite concentrate) containing about 8,300 lbs of rhenium. Part of these concentrates is exported and the other part processed locally by CARBOMET, which produces molybdic oxide, ferromolybdenum and recovers rhenium in the form of technical ammonium perrhenate.

With construction of a new by-product molybdenite plant at Chuquicamata, which will have an annual 18 to 30 million pound molybdenum output, and with construction of a molybdenite plant at Andina and expansion of the same at El Teniente and El Salvador, the molybdenum output from 14 million pounds in 1974 will increase to 27.6 million pounds in 1976, or practically double. This will increase rhenium availability from 8,500 lbs to 16,500 lbs per year of which a greater part can be recovered.

Present Chilean plans contemplate in this area two important projects; one, expansion of CARBOMET for treatment of 17 million lbs of Mo contained in molybdenite concentrates; second, construction of a new plant at San Antonio for treatment of an equivalent of 9 million lbs of Mo.

Under these plans Chile should be able to produce in a few years more:

	CARBOMET	*CAP-San Antonio*
Molybdic Oxide (Mo equiv)	17,000,000 lbs	8,600,000 lbs
Ferromolybdenum (Mo)	2,600,000 lbs	—
Ammonium molybdate (Mo)	—	400,000 lbs
Rhenium	7,700 lbs	4,000 lbs

CARBOMET will produce ammonium perrhenate while CAP will produce perrhenic acid. Only 1.6 million lbs of moly will be left for export as molybdenite concentrate.

The world rhenium production has developed in the following way:

1930	—	175 lbs
1940	—	880 lbs
1950	—	1,540 lbs
1960	—	4,400 lbs
1970	—	8,250 lbs

TABLE 6.4

Chilean Molybdenum Production and Rhenium Contents of MoS_2 Concentrates
(expressed in lbs)

Year	Chuquicamata	El Salvador	El Teniente	Total	Re content - lbs
1968	2,816,000	2,944,480	2,956,800	8,717,280	6,600
1969	4,840,000	2,200,000	3,520,000	10,560,000	7,500
1970	5,500,000	3,449,000	3,942,400	12,892,000	9,300
1971	8,198,740	2,913,240	2,854,060	13,966,040	7,700
1972	7,414,440	2,270,840	3,276,460	12,961,740	8,300
Projected					
1974	7,000,000	2,600,000	4,400,000	14,000,000	8,500
1975	12,500,000	2,600,000	5,000,000	20,600,000*	12,300
1976	18,000,000	2,600,000	6,000,000	27,600,000*	16,500
1977-80	18,000,000	2,600,000	6,000,000	27,600,000*	16,500

* These totals include expected by-product molybdenum production of Andina which should eventually reach 1,000,000 lbs per year.

However, most of this production was used not for industrial or practical purposes but for research and laboratory work. The U.S. Bureau of Mines gives the following details about rhenium production and consumption in the United States:

	Production	Consumption
1966	1,620 lbs	1,040 lbs
1967	1,725	850
1968	2,400	775
1969	3,500	2,000
1970	5,900	3,150

In recent years, rhenium consumption has been greatly increased because of new and increased uses of rhenium for catalysis. It is claimed that rhenium combined with platinum produces a stable catalyst which increases the yield and octane number of motor fuel components. Also, this bimetallic catalyst is indispensable for the production of unleaded fuel and possibly for exhaust gas control related to environmental problems. So the oil industry became the primary consumer of rhenium. In 1972 it consumed in the United States some 5,200 pounds of rhenium out of a total production of 6,000 pounds, and it is expected that in the seventies this consumption may drastically rise. The other uses of rhenium are in fabricated devices such as high-temperature thermocouples, coatings and flash bulb filament wires.

Production Technology

Rhenium recovery from molybdenite concentrates is relatively simple and consists of the following steps:
1. Roasting of molybdenite concentrates and scrubbing of gases.
2. Conditioning of the solution for maximum oxidation of rhenium values and the elimination of other impurities.
3. Recovery of rhenium values by ion-exchange and elutriation of resins.
4. Rhenium sulfide precipitation and processing into rhenium salts.

Although the first commercial production of rhenium was started in 1930 by Feit in Germany,[11] it was not until the late fifties that viable commercial processes were developed. Professor A. D. Melaven of the University of Tennessee produced the first rhenium in the United States during the Second World War, but it was Kennecott Copper

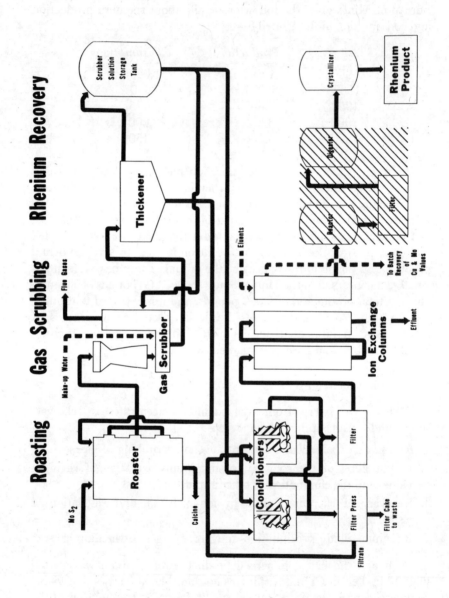

Fig. 6.4 Rhenium recovery flowsheet.

Corporation which developed the first modern process for rhenium recovery by ion-exchange techniques, largely using the basic principles developed by Melaven. This process was recently described by Prater and Platzke as follows:[12]

Molybdenite concentrates are, first, roasted in 12 hearth 19'6" Bartlett Snow Pacific multiple hearth roasters at a temperature between 540°C and 650°C, at which molybdenite is converted into molybdic oxide and about 90 percent of rhenium volatilized. The gases from the roaster are passed through cyclones and electrostatic precipitators to remove their dust burden and then go to a high-energy Ventury type scrubber for rhenium absorption along with sulfur oxides and some molybdenum. The scrubbing solution is intensively recirculated and the pressure drop in the Ventury troat is maintained in excess of 50 inches in order to maximize rhenium recovery, which in the gases reaches only a 10 to 20 ppm concentration and should be upgraded in solutions to concentrations of 0.2 to 0.5 grs/l in order to be further processed.

The pregnant solution is continuously bled off and conditioned with chlorine which oxidizes molybdenum, rhenium and iron values. Then sodium carbonate is added to precipitate iron and copper as the carbonates, and pH is adjusted to a value of 10 with sodium hydroxide. Oxidation of molybdenum must be complete because otherwise it will co-adsorb with rhenium on resins and contaminate the final rhenium product.

After precipitation the solution is filtered, transferred into a holding tank and then fed into ion exchange columns with 7 cubic feet of strong base, quarternary ammonium chloride anion exchange resins. For the adsorption cycle, two columns are used in series with 20 to 50 mesh resins. Rhenium is adsorbed selectively in this step and molybdenum and sulfate ions are rejected. When loaded, each column will contain from 25 to 30 pounds of rhenium, which is then rinsed with water to displace feed solutions and stripped with sodium hydroxide to remove any adsorbed molybdenum and with dilute hydrochloric acid to lower pH and remove any iron and copper from the resin. The rhenium is then elutriated with a 0.5 molar solution of perchloric acid. The elutriation of rhenium is counter-current.

The rhenium-rich solution is then transferred in 500 gallon batches to a 750 gallon steam jacketed glass lined reactor and acidified to 15 percent by volume with hydrochloric acid and heated to 140°F. Rhenium sulfide is then precipitated with hydrogen sulfide, filtered and washed on a pan filter.

6.5 *Ion Exchange section in rhenium recovery plant at Magna. (Photograph by Don Green, courtesy of Kennecott Research Center.)*

The rhenium sulfide is decomposed then batchwise with hydrogen peroxide and ammonium hydroxide, forming a soluble ammonium perrhenate, which is recovered by crystallization. Lately this process has been modified by using ammonium thiocyanate as an elutriant instead of perchloric acid. This permits the elimination of hydrogen sulfide precipitation, because ammonium perrhenate can be directly recovered from ammonium thiocyanate elutriant by evaporation of water and the cooling of the solution. The old and modified flowsheet can be studied in Fig. 6.4.[12]

This basic process for rhenium recovery has several modifications. At Shattuck Chemical Company in Denver, rhenium is extracted from scrubber liquor by solvent extraction rather than by ion-exchange resins. In the U.S.S.R., molybdenite concentrates are calcinated with lime, thus forming soluble calcium perrhenate and insoluble calcium molybdate. The calcines are then leached with water and rhenium recovered by precipitation with potassium chloride.

Hazen Research and Continental Ore Corporation have recently reported[14] development of a new modification for a process for rhenium recovery from molybdenite concentrates. By using a flash roasting furnace instead of multi hearth furnaces and an oxygen rather than the air atmosphere they claimed to be able to obtain higher rhenium recoveries on the order of 87 percent for concentrates containing 500 ppm of Re and more than 90 percent for premium concentrates containing 1,400 ppm Re (from Island Copper and Nevada). The key to success is apparently far lower dilution of gases due to the absence of nitrogen and additional air. In other parts the process is similar to conventional processes. A plant using this process with an annual processing capacity of 9,000,000 lbs per year is planned to be installed at San Antonio in Chile.

Investment and Production Costs

A recent U.S. Bureau of Mines publication [13] gives an interesting evaluation of capital and operating costs for rhenium production. Based on 8,000,000 and 16,000,000 pounds per year output it gives the following breakdown of costs:

	8,000,000 lbs/year	16,000,000 lbs/year
Capital Cost	$1,488,000	$2,485,000
Direct operating cost per year	356,200	599,000
Rhenium production lbs/year	3,150	6,300
Operating costs per pound	113	94

With the growing demand for rhenium and the current $1,000 plus price this investment seems to be very attractive. It is, however, related to selling molybdenite trioxide instead of molybdenite concentrate, a trend which has been increasingly resisted by industrial nations.

BIBLIOGRAPHY FOR CHAPTER SIX

1. Alexander Sutulov: Molybdenum and Rhenium Recovery from Porphyry Coppers; University of Concepcion, Chile, 1970.

2. Alexander Sutulov: An extraordinary case in primary flotation of molybdenite; Bol. Soc. Chilena de Quimica, Vol. XII, No. 2, 1962, pp. 7-10.

3. N. E. Plaksa: Steam Treatment and Selective Flotation of Copper-Molybdenite Concentrates; Tsvetnyie Metally, 1970, No. 1, pp. 79-82.

4. S. I. Mitrofanov and S. B. Dudenkova et al.: Theory and Practice of the uses of flotation reagents; Moskva, 1969.

5. A. A. Golikov and F. I. Nagirniak: Conditions of efficient depression with cyanide during selective flotation of sulphides; Tsvetnyie Metally, No. 1, 1963, pp. 5-10.

6. J. D. Vincent and W. Bossard: Mission Byproduct Plant Operation and Control: AIME Annual Meeting, New York, 1966.

7. Wayne Gould: Description of Copper Cities Concentrator, Miami, Arizona; Personal Communication, 1970.

8. Anthony Gomez, Jr.: Description of Duval's Mineral Park Operation; AIME Annual Meeting, Arizona, May 1966.

9. Rush Spedden, John Pratter et al.: Processing molybdenite concentrate by an Acid-Bake-Leach-Flotation Method; AIME Centennial Meeting, New York, 1971.

10. Alexander Sutulov: Molybdenum Extraction Metallurgy; Concepcion, Chile 1965.

11. W. Feit: Z. Angew, Chem. 43, 1930, p. 459.

12. J. D. Prater and R. N. Platzke: Extractive Metallurgy of Rhenium; AIME Meeting in New York, March 1971.

13. K. Shimamoto: Availability of Rhenium in the United States; USBM Information Circular No. 8573, 1973.

14. R. B. Coleman, A. W. Lankenau and J. E. Litz: A New Process for Recovery of Rhenium from Molybdenite Concentrates; Pacific-Southwest Mineral Industry Conference, Phoenix, Arizona, April 1973.

CHAPTER SEVEN

Smelting and Refining of Copper

Copper concentrates obtained from porphyry coppers assay generally from 20 to 50 percent copper; from 10 to 30 percent iron; and from 7 to 12 percent insoluble. They are composed of copper and iron sulfides and of gangue minerals and the purpose of smelting is to separate copper from sulfur, iron and gangue. If these concentrates are of a different nature, i.e., cement copper from precipitation plants, they undergo a separate smelting or are admixed to sulfide copper concentrates for smelting. If copper is obtained from oxide copper minerals by leaching, the final refined product is obtained without the smelting step.

The smelting metallurgy of copper is based on its strong affinity to sulfur and on its relatively weak affinity to oxygen, as compared with sulfur, iron and other base metals in the ore. Thus smelting incorporates three major steps: roasting, reverberatory furnacing and converting.

Roasting is essentially needed to eliminate the excess sulfur in concentrates in order to prepare a calcine with an ideal proportion of copper, iron and sulfur to make a good matte. It also serves for the elimination of some volatile components of the ore such as arsenic, selenium, antimony, zinc, cadmium and others.

Reverberatory furnacing has as a principal objective the elimination of gangue minerals in the form of slag and the preparation of a satisfactory matte for converting. Among the eliminated products are mostly oxygen containing gangue and a part of iron.

The third step in copper smelting, converting, is used for the oxidation of iron and the remaining sulfur, and their elimination in the form of slag and gases, thus obtaining a raw metallic copper product known as blister copper.

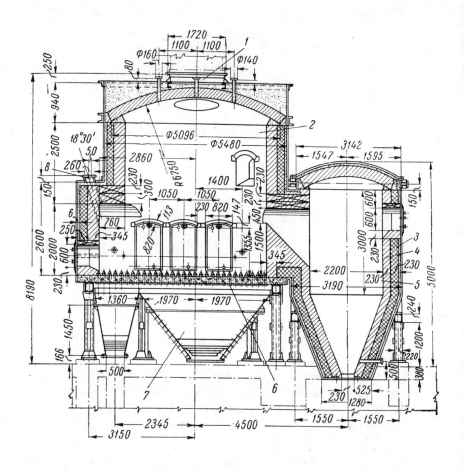

Fig. 7.1 Fluo-Solid roaster for roasting of copper concentrate pellets of Russian construction.

Roasting of Copper Concentrates [1]

Copper concentrates can be smelted in reverberatory furnaces either without or with previous roasting. Roasting is carried out only if an excess of sulfur is present in the copper concentrates. This is the case when the copper concentrates assay more than 20 percent sulfur. Since the amount of sulfur in the reverberatory furnace charge determines the grade of matte formed, the elimination of the excess sulfur helps the upgrading of mattes and the capacity of the smelter.

Roasting can be oxidizing, sulfatizing and chlorination. The last two are used for the posterior hydrometallurgical treatment of concentrates. In porphyry copper smelters, however, oxidizing roasting is used almost exclusively. Its main objective is the elimination of a part of the sulfur in the form of sulfur dioxide, the elimination of volatile components of the concentrate and partial oxidation of iron in order to convert it into slag when reverberatory smelting is undertaken.

Roasting is carried out at temperatures which are below those where the fusion of mineral constituents occurs. It is essentially an oxidation process for the efficiency of which the intimate contact between sulfide particles and oxygen should be insured. This is provided by the intensive agitation of the calcines and ready access of the air in the furnaces.

In contemporary practice roasting can be effected either in multiple hearth furnaces or fluid bed roasters. The advantages of fluid bed roasters are that they do not have very many movable parts and that they produce an SO_2 gas of high concentration, which is then more suitable for sulfuric acid fabrication. This makes the maintenance of fluo-solid roasters cheaper and provides sulfuric acid plants with 12% to 14% SO_2 rather than with 3% to 7% SO_2 from multiple-hearth roasters.[2]

The best known fluo-solid roasters are of the Dorr-Oliver and Lurgi type, but they do not have such a common distribution as in the zinc and nickel industry because of heavy investment in conventional roasters which the copper industry is reluctant to get rid of.

Because of the intensive and intimate agitation in the fluo-solid reactors the carry-over of calcines in the fluo-solid reactors is 80 percent as compared with only 6 percent in multiple-hearth furnaces. They thus require a more intensive and complete dust handling system, which consists of Cottrell precipitators.[4]

The roasting operation is normally carried out at temperatures between 620°C and 700°C which favors the formation of Fe_2O_3, which incidentally is a catalyzer for the formation of SO_3 from SO_2. At temperatures above 750°C the formation of sulfates takes place.

Fig. 7.2 New reverberatory furnace at Kennecott smelter at Garfield, Utah. (Photograph by Don Green, courtesy of Kennecott Research Center.)

Smelting of Calcines or Concentrates [2]

The next step in copper concentrates or calcines processing is smelting, which is carried out in reverberatory furnaces. At this stage individual particles are melted in a bath and two different products are formed: matte, which is an alloy of heavy metals, principally copper and iron sulfides and slag, which is an alloy of gangue minerals and metal oxides. Matte is easily melted and does not dissolve in oxide compounds which form the slag. It is also of a higher specific gravity than the slag and thus can be separated from it by decantation.

At melting temperature copper tends to form stable sulfides with sulfur — Cu_2S, regardless of oxidizing, reducing or neutral atmosphere. If the matte contains an excess of sulfur this will be bound by iron thus forming iron sufide — FeS. Both sulfides are soluble in each other in liquid form. Thus, depending on the presence of an excess of sulfur and the availability of iron, the copper matte can contain from 10 to 80 percent copper. These mattes can be melted within the range of temperatures 1130°C and 1193°C. Their specific gravity will also vary from 4.8 to 5.5, the higher being for the higher percent of copper. These mattes contain precious metals, some other sulfides and very small quantities of oxygen.[2]

If the sulfur content of the calcines is not sufficient to convert all of the iron into sulfide form, some magnetite is being formed. Magnetite has a specific gravity of 5.1 and thus is very dangerous for contamination of matte if it is poor in copper and has a lower specific gravity.[18]

To lower the temperature of melting, fluxes are added. The type of flux will depend on the composition of the concentrates and apart from common fluxes such as quartz, lime and iron ores, some other ores containing copper or gold can be added. This is to recover small quantities of copper or precious metals for which copper matte is a collector.

Smelting is normally done in 110 ft x 34 ft furnaces with a suspended basic roof. Heat is supplied either by gas or coal burners and approximately 4,000,000 Btu are necessary per ton of hot calcine to be melted. The efficiency of these furnaces is not high because the theoretical consumption should be in the range of 900,000 Btu.[10]

The furnace will smelt any feed containing from 12% to 16% sulfur and even the unroasted concentrates with less than 20 percent sulfur. During the smelting process gases and flue particles evolve from the furnace. Since the gases are hot they are passed through waste-heat boilers and then through electrostatic precipitators and

Fig. 7.3 Convertor in operation at Kennecott Smelter at Garfield, Utah. (Photograph by Don Green, courtesy of Kennecott Research Center.)

scrubbers, prior to being released in the atmosphere through a stack. These gases contain only 0.5% to 1% of SO_2 and thus are too poor for the fabrication of sulfuric acid, but are harmful from an environmental point of view.

The reverberatory furnaces are also often fed by returned convertor slag, which contains from 45 to 48 percent iron and between 1 and 5 percent copper; this is done to recover copper from slags. Also, at this point cement copper and other reject products containing copper can be fed into the reverberatory furnace.

The roasted calcines are fed into reverberatory furnaces at temperatures which vary from 400° to 600°C. The melting is done at about 1100°C to insure the melting of gangue minerals and rapid decantation of the matte. A large-sized reverberatory furnace melts about 1,500 tons of green feed per day and its cost is about $4 million.

Converting [15]

Converting is the last stage in smelting copper sulfide concentrates, which is carried out by blowing tin streams of air through a molten matte which typically contains 30 percent copper, 40 percent iron and 26 percent sulfur. In this process, first, rapid oxidation of iron occurs which results in the evolution of sulfur dioxide and the production of iron bearing slag, which is removed. When slagging is complete, molten copper sulfide, known as white metal, is blown with air to oxidize the sulfur of the white metal thus leaving metallic copper, known as blister copper. The first stage of this operation is strongly exothermic and raises the temperature of the bath to such an extent that the second stage can be carried out easily. It also consumes a considerable amount of refractory lining and flux for slag formation. Originally, when the first furnaces were built, copper conversion was accompanied by intensive consumption of refractory lining, which consisted of rammed sand and fire clay, to such an extent that converters had to be relined after each three or four operations. Later, a durable magnesia refractory brick lining was developed and silica flux began to be added so that the damage to the lining was minimized.

For more than 60 years now the Peirce-Smith sideblown converter has been the most efficient and successful furnace in copper converting.[16] Introduced in 1909 and modified to its present design, it is today a unit 13 ft x 30 ft in the form of a cylinder, installed horizontally, capable of keeping reaction temperatures at 2,000° to 2,100°F and producing some 50 tons of blister copper per batch. This furnace is

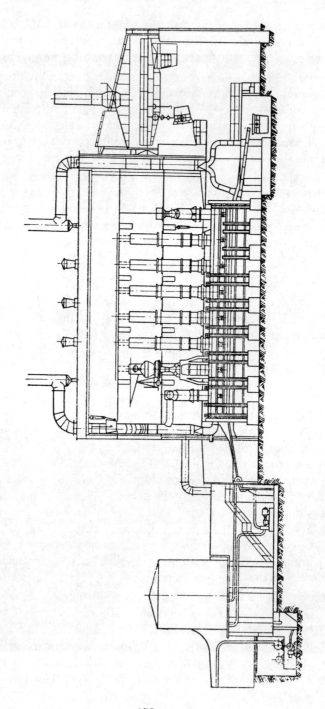

Fig. 7.4 Russian constructed electrical smelting furnace for copper concentrates.

effectively blown by air or oxygen enriched air with pipes (tuyères) submerged in matte with a blowing capacity of up to 30,000 cubic feet per minute. Typically, an 80% silica will be added for the fluxing of iron and the iron silicates skimmed off. By this process, white metal containing about 70 percent copper and 24 percent sulfur is left over with small impurities of iron, and is blown until a product of 98.5 to 99.3 percent copper is obtained. This blister copper contains about 0.3 percent sulfur, some oxygen and other impurities which are removed in a refining stage. During the converting stage a 2 to 6 percent SO_2 gas is produced at the outlet and greatly diluted by intensive blowing. This concentration may be increased up to 9% SO_2 by installing converter hoods with a tight fitting.

To produce a ton of blister, from 160,000 to 200,000 cubic feet of air should be blown through the matte. If a basic magnesite lining is used, for each ton of lining consumed some 2,000 tons of copper can be converted. Acid, silica lining consumption is at least 10 times greater.[2]

The slag formed during converting contains 2 to 5 percent copper and is either returned to the reverberatory furnaces where its high iron content is an aid to fluxing, or it is cooled, crushed and ground and subjected to a flotation operation for copper recovery.[9] When the cycle is finished, the converter is tilted to discharge the copper metal into ladles which are then transferred to anode furnaces and casting machines.

A 13 ft x 30 ft Peirce-Smith converter holds about 200 tons of matte and has a cycle of 20 hours. With a matte containing 30 percent copper it thus produces about 70 tons of blister per day. This capacity varies, of course, with the grade of the matte, and roughly 1 percent difference in the grade of the matte will vary daily output by about 2.5 tons.

Alternate Solutions

On previous pages the standard procedure for smelting copper concentrates from porphyry coppers was described. Obviously, this technology is not exclusive and has some alternate solutions. For instance, in the smelting stage low thermic efficiency and the great dilution of the SO_2 gas is criticized in reverberatory furnaces. This can be substantially corrected by electric smelting, if cheap hydroelectrical power is available as is the case in Russia and the Scandinavian countries.

Electric Smelting

In an electric furnace heat for smelting is generated by slag resistance. Thus there is no unnecessary dilution of gases and the SO_2 concentration in them rises to 2 and 4% SO_2. Also, the thermic efficiency is considerably increased because only about 400 kwh of electricity are necessary for smelting one ton of calcines.[3]

In Russia where electric smelting is used at the Dzhezkazgan, Norilsk and Alaverdy copper smelters, it has been observed that electrothermic smelting is convenient only for high-grade copper concentrates which contain relatively little sulfur and iron. If concentrates are high in sulfur, the desulfuration ratio is low and mattes are low in copper content. On the other hand, with a high iron content, copper losses in slags are not high. At Dzhezkazgan the thermic efficiency of the electric furnaces is not too high at 550 kwh consumption of electricity per ton of concentrate. It is thought that with the upgrading of concentrates to some 45% Cu it can be brought within the reach of international standards.[3]

Presently electrothermic furnaces are being built with ratings from 3,000 kva to 50,000 kva. At Severonickel Combine in Norilsk, the furnaces are of 30,000 kva. Since heating of these furnaces depends on the electrical resistance of slags, slag composition should be carefully maintained within proper limits, generally between 36 and 38 percent silica. The depth of the slag is also related to the capacity of the furnace. It is about 100 cm at Norilsk and should be about 50 percent deeper for a 50,000 kva furnace.

Flash Smelting

Another alternative for smelting is, of course, flash smelting, developed by Outokumpu Oy, but so far it has not been used on copper porphyries with the exception of one attempt at Almalyk.[4-5] At Almalyk copper concentrates containing 18% Cu, 28% Fe, 30% S and 11% SiO_2 are injected along with flux and oxygen enriched air into a furnace in suspended form where autogenous flash-combustion takes place on account of the sulfur present. A matte assaying 55% Cu, 21% Fe and 22% S is thus produced and a slag assaying 1.2% Cu, 42% Fe, 1.5% S and 23% SiO_2 is discharged. This process had some difficulties and was criticized for its not too high efficiency.

With respect to converting practices, Hoboken has produced an improved siphon-type converter which has the advantage of producing SO_2 gas of higher concentration, typically 8% SO_2, but which is more expensive capital-wise. It was recently installed at Inspiration.

Finally, there is a proposition to carry out all smelting and converting in one vessel, which should greatly decrease capital costs and increase the thermic efficiency of smelting. The most prominent alternatives so far offered and tested on pilot-plant scale are the Noranda, WOCRA and Mitsubishi processes.

Noranda Process

In the Noranda Process, concentrate and flux are fed in the form of pellets into a horizontal cylindrical furnace heated by a burner flame. Injected air agitates the mixture of slag, matte and pellets as it slowly advances to the tapping ports. Oxidizing gas, which is introduced through air tuyères, first oxidizes iron sulfide and then the remaining white metal, which is then discharged periodically as blister copper, after it separates from the matte by settling. Slag containing 9 to 12% Cu is discharged at a higher level, having been previously treated with reducing gas, also supplied through tuyères. It is then cooled, crushed, ground and floated to produce a high-grade copper concentrate which is recirculated. The tails from this flotation, which are the only reject, contain only 0.5% Cu.

An 800 tpd installation should be ready by 1973 at a cost of about $19 million. It culminates several years research and a 100 tpd pilot-plant operation. The plant also produces high-grade SO_2 gas suitable for sulfuric acid production.

WORCRA Process [16, 21]

Another extensively tested and publicized process is WORCRA (which stands for the abbreviation of its inventor, Howard WORner of Conzinc Riotinto of Australia Ltd.). In this process concentrate is fed into a central bowl-shaped smelting zone from which matte progresses to one side of the furnace where it is converted into blister copper and discharged, and the slag is moved in the opposite direction into a slag cleaning zone and discharged at the opposite end. Concentrate smelting is also added by the injection of siliceous flux and by oxygen enriched air introduced by air lances. This injection creates the necessary turbulence and continuous flow in the smelting and converting zones. In the converting zone where the white metal is rapidly converted into metallic copper, the furnace has a downward slope which permits the decantation of heavy blister into a copper well, from which

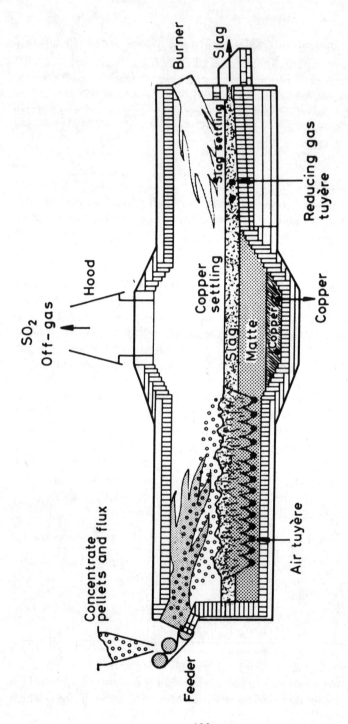

Fig. 7.5 Noranda process reactor.

it continuously overflows. On the other side and in the opposite direction, actually in a counter-current flow, slag is conducted into a cleaning zone where concentrates or pyrite are added to settle and separate entrapped matte and return it by gravity to the smelting zone via a sloping hearth.

This furnace is able to produce a 98.2% to 99.5% Cu blister with 0.6% to 0.9% S and a slag with as little as 0.3% to 0.5% Cu. Due to the use of oxygen the concentration of SO_2 in the gases ranges between 9 and 12 percent, which is ideal for a sulfuric acid or sulfur producing plant. It also claims lower operation costs and a 20 to 30% savings in capital costs as compared with conventional plants. Another advantage is that this operation can be carried out on a very small scale, say 30 to 60 tpd of copper. In fact, the pilot-plant testing done so far was on the basis of 80 tpd of concentrates. Blister from this plant should undergo fire refining prior to anode casting, because the process is not efficient in the elimination of such impurities as arsenic, antimony and others.

Mitsubishi Process

The Mitsubishi Process, although continuous in nature, is not carried on in one but in three vessels. Its basic advantages are pollution-free processing because all sulfur is eliminated in the form of 10% SO_2 gas which is easily converted into sulfuric acid or elemental sulfur, and a 30 percent reduction in investment and operation costs. The process is carried out in three units, of which the first is a smelting furnace where concentrates, flux, air and oxygen are injected through lances in the roof of the furnace. A burner with fuel oil is available if additional heat is necessary. Matte and slag are then transferred to a slag-cleaning furnace where slag is washed by pyrites mixed with coke. The relatively clean slag, containing only 0.4% Cu, is then granulated and discharged while the copper matte is sent to a converting furnace where it is converted into blister copper with the aid of coolant air introduced through the lances in the roof. The converter slag is recirculated to the smelting furnace. This converter slag runs from 7% to 15% Cu, 40 to 50% Fe and 10 to 20% SiO_2. Blister contains 98 to 99% Cu and 0.4 to 0.8% S and other impurities. The process was tested first on a 20 tpd basis and then on a 60 tpd semi-commercial basis.

Current Practices and Tendencies

There are two strong influences on current metallurgical practices in the copper smelting business leading to technological alternatives in

Fig. 7.6 ASARCO's vertical continuous cathode furnace at Chuquicamata, Chile.

development: 1) a natural desire to increase efficiency, particularly the thermal balance of smelting operations and 2) ecological considerations. It happens that smelting, principally due to the emanation of noxious SO_2 gases, is the most hazardous and polluting operation in the copper business. It is also the most visible one and thus the hardest hit by ecologists.

While concentrate roasters and converters discharge SO_2 gases of a certain concentration which can be used for the fabrication of sulfuric acid and thus the elimination of pollution problems, reverberatory furnaces discharge gases of such low concentrations that neither their conversion into sulfuric acid nor their elimination by some other means is an easy problem. The avenues of approach so far known, at least, are so costly that the use of them would be a substantial burden in additional costs. This situation is also complicated by the fact that there is no even policy of approach to this situation around the world: while the United States' regulations are extremely rigid, those of Japan and other industrial nations are less so, and in other parts of the world almost nonexistent. This, then, poses a question of unfair competition. While some industries, particularly those in the United States, have to take all the burden of ecological problems and costs, others are relatively free of these expenses and thus more competitive in production costs.

Lately, however, the situation has been turning out to be less dramatic. While only a few years ago serious thinking about relocation of smelters or the total elimination of reverberatory furnaces was undertaken, today the general tendency seems to be to replace pyrometallurgical "hot" smelting by "chemical" or hydrometallurgical conversion of copper concentrates into final metallic products, or what is even more encouraging, to develop new pyrometallurgical technology which will combine all the advantages of oxygen smelting with the autogenous smelting properties of many copper concentrates and come out with cheaper and more effective and cleaner processes.

A couple of years ago, making an excellent survey on the art of smelting,[13] Holderreed stressed the inconveniences of reverberatory furnaces and announced their possible replacement either by electrical or flash furnaces. He also criticized the conventional barrel-type converters and indicated their abandonment because of gas-control factors. He indicated then that a time would come when metallurgists would push for the elimination of sulfur as much as possible prior to roasting or flash smelting steps and come to conversion with mattes which in composition approach white metal.

Holderreed listed the following advantages in a flash smelting approach:

1. It will utilize the inherent calorific power of the iron sulfide in the concentrate as heat energy for autogenous roasting of concentrates, thus making substantial savings in direct costs.
2. Direct and indirect benefits may be expected from the diminished volume of furnace gases since the gases will require less cleaning of particulate and gaseous constituents due to the absence of fossil-fuel combustion products.
3. The SO_2 content of the furnace gases will be 8 to 12% by volume, or 15 to 25 times greater than that from conventional reverberatory furnaces. This increased content will be proportionally less costly to fix as elemental sulfur, as sulfuric acid, or an insoluble sulfate after an intermediate reaction with limestone or milk of lime.
4. The smelter will accept and successfully treat sulfide concentrates of lower copper assay than present charges.
5. Within broad limits, the roasting step in the shaft will sulfurize the iron, eliminating the subsequent slagging and removal from the skimming bay of the flash furnace, rather than postponing that chore until the converter step.
6. The smelter will release sulfur in a steady-state stream, facilitating subsequent gas processing with less interruption to the copper cycle.

The Russians, on the other hand, have been intensively researching the possibilities for clean and more efficient processes by the use of electric furnaces and oxygen blast.[3] In both cases substantial increases in SO_2 concentration were expected because of the decreased volume of diluting gases, in the first case of fossil-fuel combustion products and in the second case of nitrogen. They also came up with the so-called Kivcet Process,[7] based on cyclone-type furnaces which perform the smelting of copper concentrates in 100% pure oxygen and produce a matte which is then converted into blister in electric furnaces. The process is autogenous if the concentrates contain only 20% S and SO_2 concentration in gases varying from 70 to 85 percent. The use of oxygen, of course, intensifies considerably the processes by generating higher temperatures, sometimes posing very serious problems as was found at Bingham and El Teniente. Another fringe benefit in the use of oxygen is the substantial savings on fuels. By using flash smelting techniques at Almalyk, the Russians claim that while maintaining equal copper recoveries with the conventional reverberatory furnace and copper content of slag, they were able to reduce the heat consumption from 1,300,000 kcal/ton to about 545,000 kcal/ton; to obtain an SO_2 gas in the plus 80 percent range; to recover 90 percent of the sulfur instead of the previous 64 percent; and to increase the productivity of the furnace from the standard 4 to 5 tons per square meter of floor to about 14 tons.[5]

The use of oxygen at the Balkhash smelter has also resulted in substantial savings of fuels: an increase of the oxygen content of air to 29 percent reduced fuel consumption by approximately 33 kgs/ton (roughly 230,000 kcal) and increased SO_2 content of exit gases from 1.5 to 2.5 percent. Of course, this apparent economy in energy should be weighed against energy consumption in the production of oxygen, which runs about 170,000 to 230,000 kcal per ton in the above mentioned case.[3] Uses of oxygen in Russia, however, encountered problems similar to those at Bingham and El Teniente, particularly in the consumption of refractories. Dzhezkazgan also reported a 35 to 40 percent reduced life of refractories with the use of oxygen. The problem is particularly severe in converters, subjected to periodic heating and cooling.

In a more recent paper, White [11] observed that over the past few years new copper smelters did not adopt reverberatory furnace technology, mainly because of emissions control and improved economics. He pointed out the fact that in spite of the general use of conventional reverberatories in most of the 73 smelters around the world, there was an unmistakable trend toward the newer flash smelting, electric and continuous smelting systems in new openings, and even in old smelters some 20 companies are introducing or have already introduced novel equipment.

The Japanese copper smelting industry probably should be considered the most advanced and modern one, due principally to the postwar industrial and economic boom which permitted the construction of completely new and contemporary facilities. Besides, Japan as well as Western Europe faces severe pollution problems because of the relatively small size of her territory and notable overpopulation. The United States has a more conventional technology because of the older age of her industrial installations and because until recently pollution problems in the sparsely populated Southwest where most of the 16 U.S. smelters operate, were not severe. This situation, however, started to change drastically under the pressure of new ecological legislation.

In other parts of the world such as Canada and Australia, problems are not yet severe because of relatively small production and sparse population, but nevertheless these countries are actively working on the development of new, clean and more effective processes — Noranda in Canada and Conzinc Riotinto in Australia. On the other hand, in Africa and Latin America little progress is reported, although important copper producers such as Chile, Peru, Zambia and Zaire are quite abreast of modern technologies through international contacts, and some of them are quite willing to use more effective techniques. In Chile, for instance, at Las Ventanas, a flash smelter will be installed with Finnish help, and in Africa some smelters are switching to electric furnaces. Table 7.1 gives a complete list of smelters which process concentrates from porphyry coppers, with their respective capacities.

TABLE 7.1

SMELTERS OF PORPHYRY COPPER ORES

United States

Kennecott, Garfield, Utah	300,000
Kennecott, Hayden, Arizona	80,000
Kennecott, Hurley, New Mexico	85,000
Kennecott, McGill, Nevada	65,000
Magma, San Manuel, Arizona	200,000
Phelps Dodge, Douglas, Arizona	155,000
Phelps Dodge, Tyrone, New Mexico	100,000
Phelps Dodge, Morenci, Arizona	90,000
Phelps Dodge, Ajo, Arizona	75,000
Anaconda, Anaconda, Montana	190,000
Asarco, Hayden, Arizona	90,000
Asarco, El Paso, Texas	85,000
Inspiration, Miami, Arizona	77,000
	1,592,000

Soviet Union

USSR, Balkhash, Kazakhstan	180,000
USSR, Almalyk, Uzbeckstan	140,000
USSR, Alaverdy, Armenia	80,000
	400,000

Chile

CODELCO, Chuquicamata, Antofagasta	330,000
CODELCO, El Teniente, O'Higgins	280,000
CODELCO, El Salvador, Atacama	110,000
ENAMI, Las Ventanas, Valparaiso	60,000
Mantos Blancos, Mantos Blancos, Antofagasta	33,000
Disputada, Chagres, Valparaiso	33,000
	846,000

Canada

Noranda, Noranda, Quebec	225,000

Mexico

Minera de Cananea, Cananea, Sonora	40,000

Peru

Southern Peru, Ilo	175,000

Yugoslavia

Majdanpek, Bor	115,000

Bulgaria

Medet, Pirdop	60,000

TOTAL	3,453,000

Obviously, not all of this 3.5 million ton capacity is used exclusively for smelting of concentrates from copper porphyries, particularly if these are custom smelters, but then concentrates from British Columbia, the Philippines and many other countries in the world are smelted in Japan, where the copper industry already has a plus 1,000,000 tpy capacity.

Chemical Smelting

Another avenue of approach to copper smelting is the replacement of the pyrometallurgical operation by hydrometallurgical ones. As Holderreed stressed, hydrometallurgy will necessarily follow the same stages which are followed in pyrometallurgy: first release the sulfur, next eliminate the iron and finally produce copper as a salt or metal.

Anaconda has recently widely publicized a new hydrometallurgical process for the pollution-free recovery of copper from its sulfide concentrates from Montana, under the name of the Arbiter Process. Copper concentrates are leached by ammonia using oxygen, but without elevated temperatures or pressure. Leaching is carried out in rubber-lined tanks, followed by counter-current washing. The pregnant solution is then filtered and copper extracted by solvent extraction. Then copper is stripped from the organic solution by an acid electrolyte and won electrolytically. The process is very simple, although subject to very precise and strict controls. It requires very simple and cheap equipment and will cost probably as little as 50 percent of a conventional smelter. It also takes care of the disposal of by-product sulfates conveniently by treating them with lime, recovering ammonia and transforming the rest into gypsum. If desired, ammonium sulfate can be produced as an alternative.

Anaconda is already in the process of building a first unit of this plant which will take care of 25 percent of the sulfide concentrates produced at Butte and recover about 100 tpd of copper.

The other chemical process tried by Anaconda is the Treadwell Process, based on cyanide leaching. The basic idea of this process is to obtain a refined copper product replacing roasting and smelting operations by a leaching with sulfuric acid and conversion and electrolytical refining by the precipitation of copper cyanide and its reduction with hydrogen. In this process, concentrates are first subjected to a sulfate roast and leached with sulfuric acid. The purified solutions are then subjected to the precipitation of cuprous cyanide with cyanhydric acid and SO_2, according to the following reaction:

$$2\ CuSO_4 + SO_2 + 2\ HCN + 2\ H_2O = 2\ CuCN + 3\ H_2SO_4$$

The precipitate, after filtering and drying, is converted into metallic copper by gaseous hydrogen, liberating cyanhydric gas for recirculation. In its experimental stage this process has suffered some modifications, the concentrates being leached with calcium cyanide and calcium hydroxide. However, some difficulties developed with corrosion.

Another newly developed hydrometallurgical process is the Cymet Process, which stands for Cyprus Metallurgical Processes Corporation. In this process, finely ground copper concentrates are leached with ferric chloride. The leached copper is then electrolytically recovered in the form of a powder which is then electrorefined.

Copper Refining

Copper delivered by smelter is generally too impure for most of its applications. It contains arsenic, antimony, bismuth, lead, selenium, tellurium and iron as major impurities plus a certain amount of precious metals such as gold and silver. These impurities vary in variety and quantity depending on the type of the ore and should be removed before the copper is marketed. This removal of impurities is called refining and can be carried out in two ways: by fire and by electrolysis.

Electrolysis is a rather expensive operation and is justified when blister contains sufficient amounts of precious metals to pay for it. It also produces the premium extra pure product. Most of the primary new production of copper is converted into electrolytic form, because many electrical uses of copper and its products require this. However, a substantial part of the copper obtained in copper porphyries is sold on the market either as blister or fire refined copper. Chile, an import producer of copper and supplier of world markets, in 1971, for instance, produced a total of 728,354 metric tons of copper of which 376,909 tons were electrolytic copper, 74,711 tons fire refined and 276,734 tons blister.[24] On the other hand, Zambia, another important copper producer, converts into electrolytic copper about 88 percent of its 700,000 tpy production. The United States, the world's largest copper producer, in 1972 produced from her own sources 1,403,866 metric tons of primary copper, imported some 215,864 tons of blister copper, 35,424 tons of copper in concentrates and ores and then used scrap from domestic and imported sources for a total output of refined copper of 2,009,300 tons. Its consumption of refined copper in that year amounted to 1,944,300 tons.[25]

On a world-wide scale the 1972 copper production, expressed in metric tons, was reported to be as follows:[25]

World Mine Production	7,022,000 tons
World Smelter Production	7,332,000 tons
World Refined Production	7,996,600 tons

This basically indicates that all new copper and a substantial amount of secondary copper are invariably converted into refined product.

Refining Process

Copper refining generally undergoes three steps: first blister copper is fire refined in an anode furnace, at which stage gases and some impurities are eliminated and a compact, homogeneous product for good electrolytic refining is formed; after fire refining, anodes undergo electrolytic refining, during which all other impurities and precious metals are removed; finally, the cathodes obtained are melted again in a fuel-fired reverberatory furnace to remove the last traces of oxygen and sulfur, thus adjusting the chemical and physical properties of the final product, and giving them commercial shapes by casting.

A typical blister copper contains from 98.5 to 99.2 percent copper, 0.01 to 0.1 percent Pb, 0.01 to 0.2 percent As, 0.005 to 0.2 percent Sb, 0.005 to 0.2 percent Se, 0.01 to 0.04 percent Te, 0.0001 to 0.002 percent Bi and variable quantities of gold and silver. Gold will normally be in quantities of 1 to 2 ounces per ton (although higher contents are also possible) and silver will generally vary from 5 to 50 ounces per ton.

In the first refining step when anodes are being prepared, blister copper is molten and compressed air is blown through iron pipes into the molten mass. This causes an intense agitation and the partial oxidation of the copper. The copper oxides dissolve in the bath and react with the remaining sulfides and certain impurities such as zinc, tin, iron, arsenic and antimony. Sulfur forms SO_2 and escapes as a gas, while other metals and non-metallics form oxides and are eliminated by fluxing. Gold, silver and some other metals such as nickel and bismuth, or metalloids such as selenium and tellurium resist this fire refining and should be eliminated in the electrolytical step.

Once the oxidation of impurities is finished, the metal mass is reduced by an operation which is called polling. This consists of the insertion of green wood poles — generally of the eucalyptus tree — into the furnace door below the bath surface. The tree burns instantaneously, producing reducing gases which convert the remaining copper oxide into metallic copper. In this way the oxygen content of the anodes is reduced from 0.9 to 0.1 percent. Lately, this traditional practice is being gradually replaced by a reduction operation with reformed natural gas.

Anodes, containing between 98.5 and 99.6 percent copper and other impurities generally in the hundreds of thousands of one percent are cast in roughly 30" x 40" shapes from 1¼ to 2" thick and from 500 to 700 pounds in weight and are placed into reinforced concrete electrolytic tanks for electrolytical refining. Refining is done by placing alternately anodes and cathodes, made out of thin sheets of electrolytic copper, properly spaced, and applying direct current for the dissolution of the anodes and building up of refined copper cathodes.

Electrolytes generally contain about 200 grs/l of sulfuric acid and 45 grs/l of copper and warm up to about 140°F for optimum operation. This is also helped by the addition of sodium chloride and some organic additives such as glues, protein and other synthetic products, which insure a smooth deposition of a firm, fine-grained cathode deposit. Also, these additives take care of anode impurities by precipitating them in the form of slime and preventing growths on electrodes and shortcircuiting. All of the electrolytic cells have a bottom outlet for the removal of electrolytic slime which consists of precious metals and other impurities. Some of the impurities will not precipitate and their accumulation has a critical point. For instance, critical concentrations for nickel are 20 grs/l, for arsenic 15 grs/l and for bismuth 0.5 grs/l. These electrolytes should be purified at this accumulation stage. The slimes are treated for copper, gold, silver, selenium and tellurium recoveries. They may contain as much as 100 to 500 ounces of gold per ton, 1,500 to 10,000 ounces of silver, 25 to 35 percent copper, and variable percentages of other metals. Anode mud is generally 6 to 17 percent by weight of anode and is removed on a monthly basis.

The electrolytic cycle is designed for periods of 21 to 42 days, and after it is completed the remains of the anodes are removed, representing about 15 percent to 20 percent of their original weight, washed and scrapped for remelting in an anode furnace.

The cathodes obtained in electrolytical refining are then melted in a fuel-fired furnace and cast in refinery shapes, which should conform to certain requirements of surface quality, internal soundness and other established commercial standards. The most commonly used shapes are wirebars, ingots, ingot bars and billets.

Smelting and Refining Costs

In a recent study,[27] the U.S. Bureau of Mines referred to smelting and refining costs. This information is of particular interest, taking into consideration that production concerns are generally secretive about their costs. In this case a relatively modern smelter of 390,000 tpy copper capacity was analyzed. As is known, modern smelting technology tries to intensify smelting processes and make them more efficient. Thus for improving the thermic balance, preheated air for the blast and oxygen enriched air are used. Also, when possible, reverberatory or roasting stages are avoided. The latter can easily be avoided when the sulfur content of the concentrates is not too high. The reverberatory furnace, on the other hand, can be by-passed when the composition of the concentrates is close to the matte. In this case only a part of the concentrates undergoes reverberatory furnace smelting while the rest is added directly to converters. Such is the case of the smelter analyzed.

The smelter consists of two sections: one, materials handling, which prepares concentrates, copper scrap, anode rejects, skulls and matte

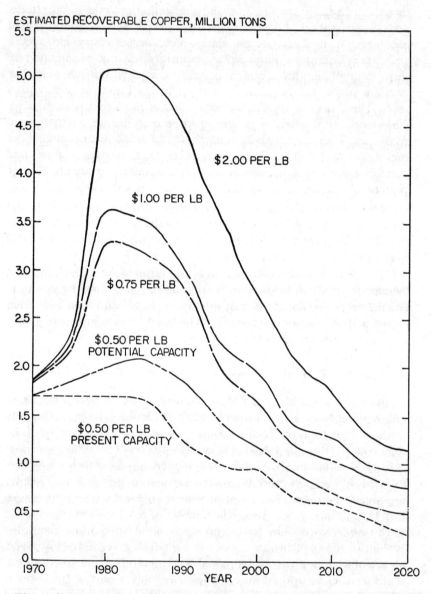

Fig. 7.7 The present and potential production capacities from U.S. resources based on price of copper (by Sheldon Wimpfen and Harold Bennett of U.S. Bureau of Mines).

shells for smelting, and the other, the proper smelting section itself. This consists of a 118 x 28 foot reverberatory furnace of 815 tons daily capacity and five 13 x 30 foot converters operated by 14,000 scfm blowers. The plant has waste heat boilers and is operated on natural gas, essentially methane. Heat consumption is about 5.6 Btu per ton of solid charge.

Concentrates, percipitates, flux and recycled dust from the converters and reverberatory gas streams and slags from the converters are fed into the reverberatory furnace to produce a 40.3% Cu matte. The matte is fed into the converters, the slag discarded and the gas and dust steam passed through waste heat boilers to produce 95,800 lbs/hr of 350 psig, 700°F steam. A part of this steam (33,000 pph) is used to preheat the air for the reverberatory furnace to 600°F, other parts (47,500 pph) for the power plant, and the balance of 15,300 pph is used in the oxygen plant.

Matte, precipitates and scrap along with other rejects are fed into the converters and 98.8% Cu blister is produced. It is then transferred into an anode furnace which produces 700 pound anodes. These are shipped to the refinery. The converter gases containing 6.7% SO_2 are channeled into a 1,075 tpd sulfuric acid plant.

The capital requirements for this plant were estimated as follows:

Materials handling section	$ 2,315,000
Smelting section	19,621,400
Oxygen plant	1,350,000
Sulfuric acid plant	9,243,000
Power plant	1,029,000
Subtotal	$33,558,400
General facilities and utilities	6,124,500
Working capital *	2,424,600
Total investment	$42,107,500

* Estimated 3 months of direct, indirect and insurance costs.

The major cost in the smelting section is for equipment — around $5 million, of which $2.2 million are for the five converters. Excavation, construction and installation work including instrumentation and insulation account for another $4.4 million, roughly half being for materials and the other half for labor. The reverberatory furnace adds approximately $3 million and the 500 foot stack and slag haulage system another $1 million. The total direct cost is $13.44 billion.

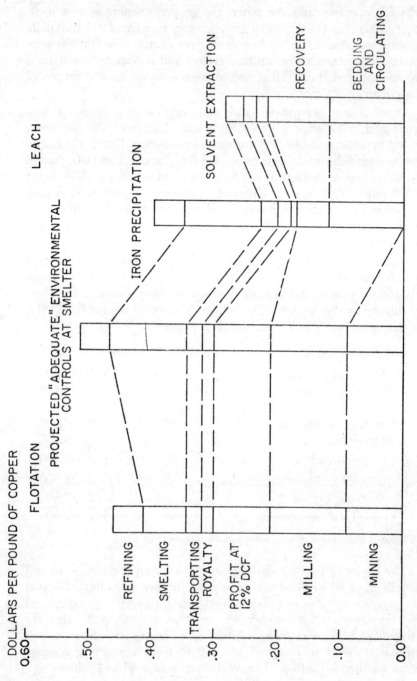

Fig. 7.8 *Comparison of estimated costs of recovery of copper by different metallurgical methods (by Sheldon Wimpfen and Harold Bennett of U.S. Bureau of Mines).*

The operating costs were based on three shifts per day, 365 days per year of operating. The credit for sulfuric acid at $3 per ton amounted to $11.8 million per year. The general breakdown of costs in dollars per pound of copper produced was as follows:

Materials and utilities	0.010
Labor	0.009
Maintenance	0.004
Overhead	0.004
Administration	0.005
Taxes and insurance	0.003
Depreciation	0.009
Gross operating cost	0.043
Sulfuric acid credit	0.004

With respect to refining, a 300,000 tpy refinery consisting of 1,566 commercial cells and four electrolyte circuits was considered. The total cost of the refinery was estimated at $33.5 million for the tankhouse, and $10.5 million for the melting and casting section with total capital requirements of $57 million when all other expenses and working capital were included.

The gross operating cost, after the deduction of by-product credit, amounted to about 3 cents per pound of copper, and allowing a 12 percent return on the investment and taxes of 50 percent on gross profit, this cost would rise to about 5 cents per pound.

Total Costs of Copper Production

Finally it would be interesting to compile the overall costs for copper production. They will, obviously, differ considerably from one country to another not only because of different costs of labor, materials and equipment, but also because of the different grade of original porphyritic ores. We already have mentioned that in Latin America, for instance, copper ores typically contain two or three times the copper of British Columbian ores, and although Latin American labor may be cheaper, the installation of imported equipment may be considerably higher in cost than in the United States where this equipment is produced.

Thus we have no alternative but to refer to the officially available statistical material. In this case the U.S. Bureau of Mines has again accomplished a very useful and interesting job by studying copper resources available in the future as a function of production costs and prices. In a recently presented paper [26] Wimpfen and Bennett show increased capacities for domestic potential in copper production as shown in Fig. 7.7. At the present average price of 50 cents a pound, the present American capacity of 1,700,000 tpy of copper can easily reach 2 million tpy in the mid-eighties, but will drop to about 1 million tons by the end of the century. However, with copper at 75 cents per pound a better than 3 million tons per year production can be achieved in the early eighties.

Returning, then, to an analysis of estimated costs for recovering copper by various methods, Wimpfen and Bennett compiled Fig. 7.7, which puts basic costs approximately in the following way, in cents per pound of metallic copper:

Mining	9 cents
Milling	11 cents
Smelting	7 cents
Refining	5 cents

In view of the information we have discussed in this book, this is a rather conservative estimate, and in other parts of the world, probably with the exception of British Columbia, copper can be produced today at less than 33 cents per pound. Now, if 12 percent profit on capital, royalties and transportation costs are included, production costs for conventionally processed copper rise to about 45 cents per pound. With projected adequate environmental control protection at smelters, this cost may increase by another 5 cents per pound. As shown in the same table, the clean copper extraction methods such as solvent extraction and cementation of copper show considerably lesser costs not only because of the lack of environmental problems but because of insignificant mining and milling costs.

Broadly, therefore, it can be concluded that porphyry coppers will continue to be the main suppliers of the new copper and that fair pricing of copper connected with increased demand and environmental protection will make even greater copper resources available for our civilization.

BIBLIOGRAPHY FOR CHAPTER SEVEN

1. V. I. Smirnov and A. I. Tikhonov: Roasting of copper concentrates; Moskva, Metallurgia, 1966.
2. L. M. Gazarian: Pyrometallurgy of Copper; Second Edition, Metallurgia, Moskva, 1965.

3. Alexander Sutulov: Copper Production in Russia; University of Concepcion, Chile, 1967.
4. L. M. Bochkarev: Fluo-Solids Smelting of Copper Concentrates; Tsvetnaya Metallurgia, No. 17, 1963.
5. L. M. Bochkarev: Flash Smelting of Almalyk Concentrates; Tsvetnaya Metallurgia, No. 6, 1965.
6. L. M. Gazarian: Voluntarism in changes of technology in copper metallurgy; Tsvetnyie Metally, No. 12, 1965, p. 11.
7. A. V. Tonkonogii: Cyclone smelting of copper concentrates; Bulletin of the Academy of Sciences, No. 6, 1956.
8. N. A. Abramov: Oxygen use in convertors of Alaverdy; Tsvetnyie Metally, No. 6, 1966, p. 22.
9. G. A. Komlev: Copper recovery from convertor slags by flotation; Tsvetnyie Metally, No. 6, 1966, p. 10.
10. Donald Treilhard: Copper — State of Art; Engineering and Mining Journal, April 1973.
11. Lane White: The Newer Technology: Where it is used and why; Engineering and Mining Journal, April 1973.
12. A. D. McMahon: Copper — A Materials Survey; U.S.B.M. Information Circular No. 8225, 1965.
13. F. L. Holderreed: Copper Smelting: Which Way in the Future?; Mining Engineering, September 1971, pp. 45-51.
14. D. MacAskill: Fluid Bed Roasting: A possible cure for copper smelter emissions; Engineering and Mining Journal, July 1973.
15. T. A. A. Quarm: Chemistry of the Copper Convertor; Mining Magazine, Vol. 117, No. 1, July 1967.
16. WORCRA Process: Smelting metals continuously; Engineering and Mining Journal, May 1967.
17. George Argall, Jr.: Copper Smelting and Refining at London Symposium; World Mining, June 1967.
18. H. W. Mossam: Magnetite in the Hurley Smelter; Journal of Metals, September 1956.
19. Seminar on Copper Reverberatory Furnace Hearth Control; Journal of Metals, February 1966.
20. Petri Bryk et al.: Flash Smelting of Copper Concentrates; Mining Engineering, June 1958.

21. H. K. Worner: WOCRA smelting-converting — a new approach to continuous direct copper production; U.S. Industrial Development Organization Meeting in Vienna, November 1967.

22. Dr. G. Leo Bailey: The application of oxygen and hot air in the modern copper industry; U.N. Meeting in Vienna, November 1967.

23. N. J. Themelis and G. C. McKerrow: Production of Copper by Noranda Process; Proceedings of the International Symposium on the Advances in Extractive Metallurgy and Refining, in London, October 1971.

24. Anuario de Estadistica Minera de Chile, para ano 1971.

25. World Metal Statistics, September 1973, World Bureau of Metal Statistics.

26. Sheldon Wimpfen and Harold Bennett: Copper Resources Apprisal; Panel III Meeting on Mineral Resources at Estes Park, Colorado, November 12, 1973.

27. U.S. Bureau of Mines Information Circular No. 8598, 1973.

INDEX

Allende, Dr. Salvador, 75, 76, 79
Alpide Belt, 20-21, 23, 27, 28, 29, 48-51, 91-95. *See also* Bulgaria; Iran; Mines in Alpide Belt; U.S.S.R.; Yugoslavia
AMAX, 31, 41, 43, 65, 67
American Smelting and Refining Co., 31, 37, 38, 41, 43, 67-69, 82, 91, 188
Anaconda, 31, 37, 41, 43, 65, 67, 82, 150, 188, 189-190
ANAMOL, 150, 155, 157
Arbiter Process, 189-190
Argall, George, Jr., 83
Argentina
  porphyry deposits in, 22, 29-30, 31, 94
  rhenium content of coppers in, 159, 160
ASARCO. See American Smelting and Refining Co.
Atlas, 48, 50, 89, 91

Bateman, A. M., 13-14
Beloglazov, K. F., 119
British Columbia. *See* Canada; Mines in Canada, British Columbia
Bulgaria
  copper production in, 18, 92, 95, 189
  porphyry deposits in, 49, 50
  see also Mines in Alpide Belt, Bulgaria
By-products derived from porphyries, 54-59. *See also* Gold; Molybdenum; Rhenium; Silver; Sulfur, separation of

Canada
  copper production in, 18, 46, 53, 54, 72, 85-87, 187-188, 189, 197, 198
  copper reserves of, 8, 10, 11, 29, 46
  molybdenum production in, 45, 54, 58, 72, 86, 87
  molybdenum reserves of, 45, 46, 54, 86, 87
  porphyry deposits in, 20, 21, 22, 28, 45, 46
  rhenium production in, 55, 86, 159, 160, 161, 163
  see also Mines in Canada
CARBOMET, 163
Chemical depression in molybdenite recovery, 148-154
Chemical smelting, 189-190
Chile
  copper production in, 18, 53, 54, 71-81, 188, 189, 191
  copper reserves of, 8, 29, 30, 31, 33, 35, 36, 40, 71, 73
  molybdenum production in, 37, 72, 163-164
  molybdenum reserves of, 30, 31, 33, 36, 71, 73
  political situation in, and copper production, 67, 71, 73, 75-79
  porphyry deposits in, 30, 31, 33-37
  rhenium production in, 160, 161, 163-164, 169
  see also Mines in Latin America, Chile
Cities Service, 41, 69
CODELCO, 31, 79, 81
Collectors, 111-115, 119-121, 145 passim
Colombia, 31, 38
Communist Bloc. *See* Alpide Belt
Conditioning reagents, 117-118
Continental Ore Corp., 41, 169
Converting, 171, 176, 177, 179, 185, 195
Conzinc Rio-Tinto, 89, 181, 188
Copper
  content of, in typical porphyry, 14, 15-16
  demand for, 7-8, 11

201

depression of, in molybdenite separation, 144-154, 155
percentage of, mined from porphyry, 10, 53
precipitation of, 61, 124-129, 156
production of
  cost factors in, 9-11, 25, 132-135, 194, 197-198
  technical processes in (*see* Crushing; Flotation; Grinding; Leaching, Refining; Smelting)
  world figures on, 10, 53-54, 191
  see also names of individual countries, copper production in
reserves of
  in Alpide Belt, 8, 10, 29, 49, 50, 93
  in Pacific Fire Belt, 10, 29, 50
  world, 7-11, 27, 29, 54
  see also names of individual countries, copper reserves of
yield of, per ton of ore, 24, 25, 55, 57, 58
Copper concentrates, cleaner, composition of, 122, 141
Copper porphyries. *See* Porphyry deposits
Cox, Dennis P., 19
Crushing and grinding, 99-105, 121-122, 133
  equipment for, modern, 101, 102, 103, 105, 133
Cymet Process, 190
Cyprus Mines, 41, 43

Dual Process, 127, 129
Duval (Pennzoil), 41, 43, 63-64

Ecuador, 31, 38
Electric smelting, 178, 179-180, 186, 187
Electrolysis, copper recovery by, 131, 191, 192-193
El Paso Natural Gas, 69-70

Fire refining, 191, 192
Flash smelting, 180-181, 186, 187
Flotation
  cleaner, 121-122, 141
  cost factors in, 107, 110
  depression of iron in, 105, 110-111, 112, 113, 117, 118, 119, 120, 126, 129
  flowsheet for, 104, 105
  modern equipment for, 105-107, 110, 133
  of molybdenite, 105, 108-109, 110,
111, 113, 115, 117, 118-119, 121-122, 141, 144-145, 154-157
  oxidation problems and, 122-127
  recovery from, at all operating mines, table, 108-109
  rougher, 110-121
Freeport Sulphur Co., 50, 89
Frei, Eduardo, 75
Frothers, 115, 117, 155

Gangue minerals
  content of, in porphyry, 17, 119
  separation of, in milling, 99, 126, 141-142, 145, 147, 155-156
  separation of, in smelting, 171, 175, 177
Gold, produced from porphyry, 53, 55, 57, 58, 87, 88, 89, 91, 133, 192-193

Hecla, 41, 43, 69-70
Holderreed, F. L., 185-186
Hydrometallurgical treatment of oxidized ores, 127-130

Inspiration, 41, 188
Ion-exchange techniques, 166-169
Iran, 23, 49, 50
Iron
  content of, in cleaner concentrates, 141, 157, 171
  content of, in porphyry, 14, 15-17, 35-36, 120
  depression of, in cleaner concentrates, 148-154
  depression of, in rougher concentrates, 105, 110-111, 112, 113, 117, 118, 119, 120, 126, 129, 130, 131-132
  elimination of, in refining, 191, 192
  elimination of, in smelting, 171, 173, 175, 177, 179, 180, 181, 183, 186
  use of, to precipitate copper, 61, 124-126, 127, 129, 156

Jackling, Daniel C., 9-10, 17, 101
Japan, 38, 86, 87, 89, 187, 189

Kennecott Copper Corp., 30, 31, 40, 41, 48, 188
  processes developed by, 129, 149-150, 156, 165, 167
  production data for, 55-59, 60, 160
Kivcet Process, 186

Latin America, 8, 10-11, 22-23, 29, 31-32, 72, 159, 160, 161, 197. *See*

also Argentina; Chile; Colombia;
Ecuador; Mexico; Mines in Latin
America; Panama; Peru; Puerto
Rico
Leaching, 157, 190
of oxidized ores, 61, 124-125, 127,
129-130
Lepanto Consolidated Mining Co., 48,
91
Lowell, J. D., 8, 9, 14, 21, 23, 45
and J. M. Guilbert, 14-15, 40
LPF Process, 124-127

McKinstry, H. E., 13
Magma Copper Co., 40, 41, 61, 63, 188
Malouf, E. E., 130
Marcopper, 48, 50, 91
Marinduque, 48, 50, 91
Mexico
copper production in, 18, 67, 189
copper reserves of, 8, 29, 31, 39-40
porphyry deposits in, 31, 39-40
see also Mines in Latin America,
Mexico
Milling. *See* Crushing and grinding;
Flotation
Mineralization and reagents, 112,
119-121
Minero Peru, 31, 37, 38, 83
Mines, tabulated data on
directory of world, 31, 41, 46, 50
flotation reagents used in, 112
metallurgy of operating, 108-109
molybdenite recovery in, 146
production data on operating, 60, 72,
88, 92
rhenium content of porphyries at, 160
with smelters, 188-189
tonnage-grade relationship in, 24
Mines in Alpide Belt, 48-51
Bulgaria
Medet, 49, 95, 123, 151
U.S.S.R.
Almalyk, 18, 49, 93, 95, 126, 151,
159, 180, 186
Balkhash, 93, 118, 120, 124, 147,
151, 187
Bozshchkul, 49, 93
Kadzharan, 49, 95, 123, 151
Yugoslavia
Majdanpek, 18, 49, 95
See also Mines, tabulated data on
Mines in Canada, 45, 46
British Columbia
Bethlehem, 23, 45, 85, 86, 94, 102
Brenda, 18, 25, 45, 55, 64, 87, 101,
115, 123, 157, 159

Gibraltar, 18, 45, 54, 85, 101, 102
Island Copper, 18, 54, 85, 101, 102,
103, 156, 159, 169
Lornex, 18, 23, 45, 54, 85-86, 101,
102, 103, 107, 135
Quebec
Gaspe, 85, 87, 147
See also Mines, tabulated data on
Mines in Latin America, 29-41
Chile
Andina, 75, 76, 77, 79, 101, 102,
159, 163
Chuquicamata, 17, 25, 33-35, 67, 74,
76, 77, 79, 81, 123, 127, 129,
141, 144, 150, 155, 157, 159,
163-164
El Abra, 36, 81
El Rio Blanco, 29, 33
El Salvador, 33, 67, 76, 77, 79, 123,
141, 150, 155, 157, 159, 163-164
El Teniente, 17, 29, 30, 59, 75, 76,
77, 79, 80, 81, 101, 102, 103,
110, 111, 115, 118, 121, 123,
141, 144, 145, 150, 155, 159,
163-164, 186, 187
Exotica, 36, 75, 76, 77, 78, 79, 131
La Disputada, 29, 30, 159
Los Pelambres, 36, 81
Mantos Blancos, 25, 33
Mexico
Cananea, 18, 39, 67, 159
La Caridad, 39-40
Peru
Cerro Verde, 37, 38, 82, 83, 94
Cuajone, 37-38, 82, 83, 94
Michiquillay, 38, 82, 83
Quellaveco, 37, 82, 83
Toquepala, 18, 37, 82-83, 159
See also Mines, tabulated data on
Mines in Pacific Fire Belt, 45, 47-48, 50
New Guinea
Bougainville, 18, 21, 23, 25, 48,
87-89, 101, 102, 103, 115
Ertsberg, 25, 48, 89
Philippines
Atlas, 18, 89, 91
Biga, 48, 89
Labo, 48, 91
Santo Thomas, 18, 48, 91
Sipalay, 48, 91
Toledo, 18, 48, 89
See also Mines, tabulated data on
Mines in U.S., 40-45
Arizona
Bagdad, 18, 59, 61, 70, 94, 115,
118, 123, 131, 132, 147, 150, 155
Copper Cities, 17, 18, 61, 69, 141

Esperanza, 63, 101, 103, 123, 132, 147, 157
Inspiration, 18, 61, 69, 70, 118, 127, 129, 132, 147, 150, 155, 157
Metcafe, 43, 59, 70, 94
Miami, 18, 59, 61, 69, 94, 123, 124, 145, 150, 154, 155
Mission, 43, 67-69, 101, 115, 118, 135, 141, 145, 147, 150, 151, 153, 155
Morenci, 17, 18, 25, 43, 70, 118, 127, 141, 156
Pima, 43, 54, 59, 65, 67, 70, 101, 102, 103, 115, 123, 135, 141, 145, 153, 157
Pinto Valley, 43, 59, 69, 94
Ray, 61, 115, 118, 123, 126, 147, 150, 155
San Manuel, 18, 25, 40, 42, 59, 61-62, 123, 147, 155, 159
Sierrita, 43, 54, 59, 63-64, 101, 102, 103, 115, 135, 141, 159
Silver Bell, 18, 69, 115, 135, 141, 145
Twin Buttes, 43, 54, 59, 65, 94, 101, 102, 123, 135, 141, 145
Montana
Butte, 43, 61, 101, 115, 126
Nevada
McGill, 18, 115, 123, 141, 147, 150, 159
Yerington, 18, 61, 67, 127
New Mexico
Chino, 18, 61, 118, 123, 147, 150
Tyrone, 59, 70, 115, 135
Utah
Bingham, 10, 18, 25, 40, 129, 186, 187
Magna and Arthur, 59, 61, 115, 123
See also Mines, tabulated data on
Mines to be developed, 83, 94, 95
Mitrofanov, S. I., and Kurochkina, 118, 152
Mitsubishi Process, 183
Molybdenite
occurrences and properties of, 54, 142-144
recovery of
chemical depression in, 148-154
flotation in (see Flotation, of molybdenite)
flowsheet for, 140, 144-145
leaching in, 157
thermal depression in, 145-148
world figures on, 159-161
rhenium recovery from, 157-170 (see also Rhenium)

Molybdenum
percentage of, in typical porphyry, 14, 15
production of (see names of individual countries, molybdenum production in)
reserves of
in Pacific Fire Belt, 50, 89
world, 54-55
see also names of individual countries, molybdenum reserves of
yield of, per ton of ore, 57, 58
Morenci Process, 148-149

New Guinea, 18, 48, 87-89. See also Mines in Pacific Fire Belt, New Guinea
Nokes reagent, 149-151
Noranda, 46, 87, 188
Noranda Process, 181, 182

Open pit mining, predominance of, 61, 99
Ores, low-grade. See Copper, production of, technical processes in; Porphyry deposits
Oxidation
flotation processes to combat, 122-127
leaching to combat, 61, 124-125, 127, 129-130
of porphyry orebody, 15-17

Pacific Belt, 20-21, 27, 28. See also Pacific Fire Belt
Pacific Fire Belt, 10, 11, 29, 45, 47-48, 50, 87-91. See also Mines in Pacific Fire Belt; New Guinea; Philippines
Panama, 25, 31, 38-39, 94
Parson, A. B., 13-14
Peru
copper production in, 18, 53, 54, 82-83, 84, 189
copper reserves of, 8, 29, 31, 37, 38, 82
molybdenum production and reserves in, 31, 37, 72, 82, 83
porphyry deposits in, 22, 23, 31, 37-38
see also Mines in Latin America, Peru
Phelps Dodge, 37, 41, 43, 70, 188
Philex, 48, 50, 91
Philippines
copper production in, 18, 88, 89, 91
porphyry deposits in, 21, 23-24, 48
see also Mines in Pacific Fire Belt, Philippines

Placer Development, 46, 86
Pollution problems in smelting, 70-71, 175-176, 183, 185-188, 189-190
Porphyries. See Porphyry deposits
Porphyry deposits
　age of various, 21-23, 25
　defined, 13-14
　distribution of, 19-21, 27-29, 32, 42, 44, 47 (see also names of individual countries, porphyry deposits in)
　exploitation of, to meet expanding demand, 8-11
　metals derived from, 53-59
　mining of, historical, 9-10, 17-18
　model of typical, 14-17
　origin of, 18-20
　rhenium content of various, 55, 65, 157, 159-161
　tonnage-grade relationship in, 24, 25, 55, 57, 58
　see also Mines, tabulated data on; Mines in Alpide Belt; Mines in Canada; Mines in Latin America; Mines in Pacific Fire Belt; Mines in U.S.
Puerto Rico, 31, 39
Pyrites. See Iron

Refining, 190-193, 197, 198
Regrinding of rougher concentrates, 121-122
Reverberatory furnacing, 171, 174-175, 177, 185, 186, 187, 191, 193, 195
Rhenium
　content of, in various copper porphyries, 55, 65, 157, 159-161
　production of, world, 57, 93, 160-165
　recovery of
　　cost factors in, 159, 160, 162, 169-170
　　technology of, 140, 165-169
　resources of, world, 159-161
　uses and consumption of, 55, 165
Rio Algom Mines, 45, 46, 87
Roasting
　in molybdenite separation, 145, 146, 147
　in smelting, 171, 172-173, 193
Russian sulfidization process, 151-153

Shattuck Chemical Co., 169
Shoemaker, R. S., and Taylor, A. D. 107
Sillitoe, Richard H., 19-21, 27
Silver, produced from porphyry, 53, 55, 57, 58, 87, 88, 89, 91, 133, 192-193
Smelters of porphyry copper ores, table, 188-189
Smelting
　chemical, 189-190
　converting in, 171, 176, 177, 179, 185, 195
　cost factors in, 185, 193-197, 198
　electric, 178, 179-180
　flash, 180-181
　Mitsubishi Process in, 183
　modern practices to improve, 183-188, 193
　Noranda Process in, 181, 182
　pollution factors in, 70-71, 175-176, 183, 185-188, 189-190
　reverberatory furnacing in, 171, 174-175, 177, 185, 186, 187, 191, 193, 195
　roasting in, 171, 172-173, 193
　WORCRA Process in, 181, 183
Solvent extraction process, 131-132
Spedden, H. R., E. E. Malouf and J. Davis, 129-130
Sulfides
　depression of, 148-157
　flotation and leaching of, 119, 123, 124-126, 129, 130, 156
　in porphyry orebody, 15-17
　rhenium, 167-168
　sodium, use of, 118, 147, 148, 151-154, 156
Sulfidization processes, 151-154
Sulfur, separation of
　in refining, 192
　in smelting, 61, 171, 173, 175, 177, 179, 180, 181, 183, 185, 186, 187
Sulfuric acid
　recovery of, from smelter gases, 69-70, 87, 126
　uses of, 69-70, 124-127, 147, 156
Sumitomo Group, 85

Thermal depression in molybdenite recovery, 145-148
Tonnage-grade relationship in porphyries, 24, 25, 55, 57, 58
Treadwell Process, 190

U.S.A.
　copper production in, 9-10, 17-18, 53-54, 59-71, 99, 191, 194, 198
　copper reserves of, 8, 40, 41, 43
　molybdenum production in, 54, 57, 58, 59, 60, 63, 65, 69
　molybdenum reserves of, 30, 41, 54
　porphyry deposits in, 20-21, 22-23, 28, 40-45
　rhenium production in, 55, 57, 159,

160, 161, 165
see also Mines in U.S.
U.S.S.R.
  copper production in, 18, 53-54, 92, 93, 95, 99
  copper reserves of, 8, 10, 49, 50
  flotation techniques in, 113, 115-117, 118, 120, 123, 124, 125, 126, 155, 156
  molybdenite recovery in, 145-148, 151-153, 155, 156
  molybdenum production and reserves in, 49, 50, 92, 95
  porphyry deposits in, 10, 18, 48-51
  rhenium production in, 159, 160, 161, 169
  smelting techniques in, 180, 186-187, 188
  sulfidization process in, 151-153
Utah Mines, Ltd., 46, 86

Western sulfidization process, 153-154
Wimpfen, Sheldon, and Harold Bennett, 198
WORCRA Process, 181, 183

Yugoslavia
  copper production in, 18, 92, 94, 95, 189
  porphyry deposits in, 18, 49, 50, 95

Zaire, 49, 54, 188
Zambia, 49, 54, 188, 191